Heinz Engelhardt
Wolfgang Beck
Thomas Schmitt

Capillary Electrophoresis

Methods and Potentials

Heinz Engelhardt
Wolfgang Beck
Thomas Schmitt

Capillary Electrophoresis

Methods and Potentials

Translated from the German
by George Gutnikov

Title of the German edition:
Kapillarelektrophorese
© Friedr. Vieweg & Sohn Verlagsgesellschaft mbH, 1994

ISBN 978-3-642-85856-7   ISBN 978-3-642-85854-3 (eBook)
DOI 10.1007/978-3-642-85854-3

Vieweg is a subsidiary company of Bertelsmann Professional Information.

Produced by Lengericher Handelsdruckerei, Lengerich

TABLE OF CONTENTS

PREFACE

Capillary electrophoresis (CE), also designated by the acronym HPCE for "High Performance Capillary Electrophoresis" is a rapidly growing analytical separation method. It unites the separation technique of classical electrophoresis on plates with the instrumental methods of chromatography with respect to direct detection of the solutes separated in the capillary and their ready identification and quantification. The initial problems of inadequate reproducibility in quantitative analysis, due to the necessity of handling extremely small volumes, have largely been solved in the second generation commercial instruments. Hence, a rapid and reliable separation system is available for ionic compounds from the smallest cation (the lithium ion) up to polyanions with molecular weights ranging in the millions (such as DNA molecules). The methods of gel electrophoresis and isoelectric focusing can be readily extended to separation techniques carried out in a capillary. For nonionic compounds an additional separation method is available in the form of micellar electrokinetic chromatography (MEKC). This involves a true chromatographic separation process because the distribution of the analytes between the buffer and the micelles is superimposed on the electrophoretic migration, which contributes substantially to the selectivity.

This book is intended as an introduction to the principles and potential applications of capillary electrophoresis. Since it originated as a result of presentations given in short courses for chemists in industry, greater emphasis was placed on the experimental aspects rather than the theoretical principles. The objective was to introduce the readers to the technique of capillary electrophoresis to the extent of enabling them to develop separation methods on their own. Since already reported separations can be readily researched via data banks, such compilations were intentionally omitted here. Instead, an attempt was made to provide a guide to troubleshooting, which surely is not complete yet, but which summarizes many of the problems that can drive the beginner as well as the seasoned practioner to despair.

This book is the result of fruitful collaboration with both coauthors as well as the assistance of my other Master's and Ph. D. candidates and the inclusion of their as yet unpublished data. These students of the CE group are Jörg Kohr, Miguel Cunat-Walter, Stefan Adam, and Christof Finkler. Here, I also wish to express my appreciation to them for their enthusiasm. My friends and colleagues Fritz Erni (Sandoz

AG, Basel) and B. L. Karger (Barnett Institute, Northeastern University, Boston), I thank for the opportunities of sending my collaborators to their laboratories. The various instrument manufacurers I thank for the temporary loan of instruments, for with laboratory-built apparatus alone the potential speed of the separations cannot be fully realized.

Saarbrücken, March, 1994

Heinz Engelhardt

PREFACE TO THE ENGLISH EDITION

I am happy that soon after the appeerence of the book on capillary electrophoresis in Germany an English edition is published. It gave me the chance to eliminate some mistakes and to add some additional and new results. I appreciated again the cooperation with George Gutnikov during his work on the translation.

Saarbrücken, July 1995

Heinz Engelhardt

1 INTRODUCTION

Under the influence of an applied electric field, charged particles in solution migrate at different velocities and in different directions. These ionic migrations were investigated about 100 years ago by Kohlrausch, who also described the physical laws governing their behavior. The concepts of "electrophoresis" and "electrical transference" were coined early on. The charge of the ions determines their direction of migration. Differences in the migration velocities can have two causes: different magnitudes of charge result in different electrophoretic mobilities in an electric field, or ions resist electrophoretic migration through different frictional resistances due to differences in size for the same effective charge. This shows the numerous separation possibilities of electrophoretic processes. Their practical introduction as an analytical method became possible when the distortion of the zone profiles, caused by convection currents that result from the dissipation of the Joule heating, was reduced or eliminated. This was achieved through the use of stabilizing gels that were poured onto a glass plate (gel electrophoresis), or by means of paper strips that were immersed in a buffer. After introduction of gel electrophoresis by Tiselius, the method developed into the most widely disseminated method in biochemistry. The use of gels also permitted the successful separation of macromolecules with very small differences in their charge densities. The separation effectiveness of the gel can be optimized by changing the degree of crosslinking. Besides DNA molecules, gel electrophoresis is used primarily for the separation of peptides and proteins. These are separated either in their native form according to their charge or, after denaturation with SDS (sodium dodecyl sulfate), according to their molecular weight in relation to their molecular size. This method also serves for the estimation of the molecular weights of proteins in solution. The separation is based on the fact that molecules of different sizes exhibit varying degrees of friction with the gel fibers, thus enabling resolution according to molecular size. Without the presence of a separating gel, all proteins denatured with SDS would migrate at the same velocity.

In flat-bed electrophoresis several separations can be carried out simultaneously on parallel lanes. Moreover, two-dimensional operation is possible in which extraordinarily high separation efficiencies can be achieved by the use of two different separation principles. The 2-D-gel electrophoresis of proteins with separation in one direction according to effective charge and in the second, according to molecular weight after denaturation with SDS, is mentioned here only as an example.

Despite the very good separation results attainable, classical gel electrophoresis has some decided disadvantages. Reproducible preparation and handling of gels poured onto glass plates is difficult. Joule heating along the separation path increases with the square of the voltage, thereby making effective cooling of the separation medium indispensable to prevent the gel from drying out during analysis. Lower field strengths lead to smaller current flows, but the analysis time increases greatly. The principal disadvantage of classical electrophoresis, however, lay in the inability to develop instrumentation to perform the separation and to detect the sample components directly, as in chromatography. The problems with the detection of proteins after their visualization and quantitative determination on the gel layer are mentioned only as an example.

Therefore, efforts were made to circumvent the problems with convection currents in gels and to realize the direct photometric determination of the sample components in the buffer. In narrow tubing the convection currents are relatively small due to its small inner diameter. The large surface area of the capillary relative to the buffer volume permits rapid, effective removal of the Joule heating and hence the use of relatively high potentials, thus leading to fast analyses. The electrolyte can be easily changed. Suitable capillary material provides optical transparency and enables direct (on-line) detection and quantitative determination of the solutes after separation, as in chromatography.

The first separation in an open glass tube was described by Hjerten [1], who achieved stabilization of the solution by rotating the tube about its axis. The actual development of capillary electrophoresis (CE) began with the pioneering work of Mikkers and Everaerts [2,3] toward the end of the 1970's and of Jorgenson and Lukacs [4,5] at the beginning of the 1980's. Advances in the method resulted from two decisive improvements: First, the inner diameter of the separation capillary was reduced substantially, and, second, the conductivity detection that had originally been adopted from isotachophoresis was replaced by on-line UV-detection. A requirement for further development was the availability of fused silica (quartz) capillaries of high transparency in the lower UV-region and uniform inner diameters in the range of 50 to 100 μm. This improved the separation efficiency as well as the detection options considerably.

High efficiency separations of proteins and dansylated amino acids could be achieved with fused silica capillaries of 50-100 μm i.d., in which thermally induced convection was greatly reduced because of the relatively large surface area to volume ratio. Fused silica capillaries permitted the use of modified detectors from HPLC for sensing the separated components directly in the capillary. The simplicity of the apparatus and the growing need for separations of biomolecules generated increasing interest in the second half of the 80's.

2 PRINCIPLES
OF CAPILLARY ELECTROPHORESIS

The concept of capillary electrophoresis encompasses various separation techniques. In capillary zone electrophoresis (CZE), which is carried out exclusively with electrolyte-filled capillaries, separation is based on the mobility differences of the solutes, and at present represents the most frequently used method of capillary electrophoresis. The method resembles elution chromatography in that the zones migrate at different velocities through the separation system and, in the optimum case, are separated from each other by buffer electrolyte.

In capillary gel electrophoresis (CGE) the capillary is filled with a polymer solution or a gel. Electrophoretic migration of macromolecules is hindered by the gel matrix. The transport of the solutes through the capillary is based on the charge of the macromolecules, but the separation is dependent on the molecular size.

Uncharged molecules can be separated by means of micellar electrokinetic chromatography (MEKC). Detergents are added to the buffer and the neutral molecules distribute themselves between the buffer and the micelles according to their hydrophobicity. Separation is based on the mobility of the most negatively charged micelles and the transport of the solutes outside the micelles via the electroosmotic flow (EOF). This involves a partition process so that one is dealing with a real chromatographic method. In this case, too, the solutes are separated from each other by electrolyte. The range of separation lies between the compounds that do not reside within the micelles and hence migrate with the EOF, and those that are permanently enclosed within the micelles.

In isoelectric focusing (IEF) separation occurs in a pH gradient formed by the addition of ampholytes to the buffer in an electric field. Electrochromatography (EC), in which HPLC stationary phases are used and eluent flow and sample transport are effected solely through the EOF, has attained less significance up to now. Finally, isotachophoresis is presented as the oldest capillary technique and has recently been gaining some importance for sample concentration in CZE. Here the solutes migrate with the same velocity and are not separated from each other by pure electrolyte. This method is analogous to displacement chromatography. The separation techniques of CE are summarized in Table 2-1, which also indicates their most important areas of application. The individual techniques will be discussed in detail in order of their importance and extent of application.

Table 2-1 Summary of separation systems in capillaries driven by electric fields

Capillary zone electrophoresis (CZE)	Micellar electrokinetic chromatography (MEKC)	Capillary gel electrophoresis (CGE)	Isoelectric focusing (IEF)	Isotachophoresis in capillaries (ITP)	Electro-chromatography (EC)
Capillaries modified/ unmodified	Capillaries unmodified	Capillaries modified and filled	Capillaries modified/ unmodified	Capillaries unmodified	Capillaries packed
EOF controlled/ uncontrolled	EOF present	EOF suppressed	EOF suppressed/ weakly present	EOF present	EOF present
Separation by electrophoretic migration, influence of EOF	Separation by electrophoretic migration and partition of sample between micelle and buffer, influence of EOF	Separation by electrophoretic migration with sieving effect, no influence from EOF	Separation in a pH gradient according to the isoelectric points of the sample. When EOF is present, mobilization of the sample zone to the detector not needed	Separation by electrophoretic migration, influence of EOF	Separation by interaction with a stationary phase (LC separation system) EOF serves as pump
Many possible applications for small and large charged molecules	Extensive application range for small, neutral, and charged molecules	Separation according to molecular size for DNA molecules and SDS-denatured proteins	Separation of zwitterionic analytes according to their pI values	Range of applications limited. Enrichment of diluted solutions prior to CE	Range of applications as in HPLC (of little practical significance)

The moving boundary method is only of theoretical interest because it does not involve a real separation method. Only the most rapidly migrating ions are separated from the other zones which are only partially separated. The migration profile corresponds to the elution profile of frontal analysis in chromatography.

A schematic drawing of a CE apparatus is presented in Fig. 2-1. A fine quartz capillary (25-100 µm i.d.) of 20 to 100 cm long bridges the two buffer containers between which a potential of up to 30,000 volts is applied. For simplicity we will always refer to quartz capillaries, even though it is actually amorphous silicon dioxide (fused silica) from which the capillaries are drawn. A relatively short sample plug (a few nL) is introduced at the anodic end of the capillary. This is performed either by lifting or lowering of the corresponding containers, through pressure on the sample container, vacuum on the buffer reservoir, or simply by electrophoretic migration of the sample components into the capillary. The advantages and disadvantages of each method of sample introduction will be discussed in detail in the Instrumentation chapter.

The separation of the sample components in the buffer within the capillary is achieved by applying a potential across the buffer containers. The electric field resulting in the capillary effects migration of the solutes. Always superimposed onto the electrophoretic migration is a more or less strong electroosmotic flow that contributes passively to the transport of the solute zones, but not to their separation. This EOF depends strongly on the pH value of the buffer and the surface properties of the capillary. It can be so large that not only neutral molecules, but even negative ions, can be transported to the detector counter to their electrophoretic migration.

In most buffers, negative charges that exist on the surface of the quartz capillary from the dissociation of the surface silanol groups induce a positive charge in the so-

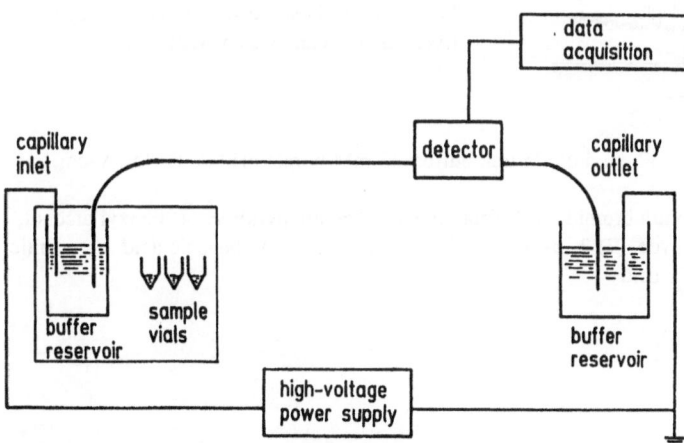

Fig. 2-1 Schematic of a CE apparatus [6]

lution adjacent to the wall. As a result, the EOF moves in the direction of the cathode, and therefore in the usual instrumental configuration the detector is located in the vicinity of the cathode compartment. The EOF helps to transport the solute zones to the detector, so that for a sufficiently high EOF anions are also transported to the cathode. A separation of cationic, anionic, and neutral substances by CZE is shown in Fig. 2-2. Under these conditions all uncharged molecules migrate with the same velocity (that of the EOF) and cannot be separated from each other, whereas the ions can be separated because their electrophoretic mobilities differ.

Beside this simplest form of CZE, a multiplicity of variants has been introduced that will be discussed later in separate chapters, both theoretically and with examples of typical applications.

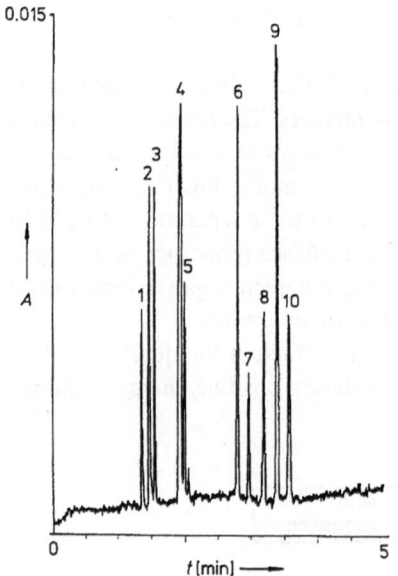

Fig. 2-2
Separation of positively, neutral, and negatively charged analytes by CZE [6]

Separation conditions: L=30/37 cm, i.d.=75 µm, buffer: 33 mM borate pH 9.5, E=350 V/cm, detection: UV/214 nm
(1) Trimethylphenylammonium bromide, (2) histamine, (3) 4-aminopyridine, (4) benzyl alcohol, (5) phenol, (6) syringealdehyde, (7) 2-(p-hydroxyphenyl)-acetic acid (8) benzoic acid, (9) vanilic acid, (10) p-hydroxybenzoic acid

3 THEORETICAL FOUNDATIONS AND THEIR INFLUENCE ON THE ANALYTICAL RESULTS

3.1 Electrophoretic Migration

Ions move at constant velocity in an electrical field. Increasing the potential and, consequently, the field strength E always raises the migration velocity u of the ions and leads to faster analyses. The electrophoretic mobility μ of the ions relates the electrophoretic velocity u and the field strength E as

$$u = \mu \cdot E = \frac{L_{eff}}{t_m}$$

(3.1)

The velocity of an ion is determined by dividing the traversed path length (the capillary length from inlet to the detector) by the migration time t_m of the peak.

This formula can be derived from the force equilibria to which a migrating ion is subjected. A single ion is subject to an accelerating force F_A

$$F_A = z \cdot F \cdot E / N_A,$$

(3.2)

where F is the Faraday constant (96,500 Cb mol^{-1}), N_A is Avogadro's number, and z the effective ionic charge. This force is opposed by the frictional force F_F, approximated by Stokes' Law

$$F_F = 6 \cdot \pi \cdot \eta \cdot r \cdot u$$

(3.3)

Here η is the dynamic viscosity [Pa s] and r is the Stokes radius [cm] of the ion, including the solvation shell. The migration velocities cannot be calculated directly using values specific to the molecules and the buffer because the Stokes radii are frequently unknown and do not correlate with the ionic radii in crystal lattices. The number 6 holds for the assumption of spherical particles, but is smaller for small ions whose size corresponds to that of the solvent molecules.

Accordingly, the electrophoretic migration velocity u is given by

$$u = \frac{z \cdot F \cdot E}{6 \cdot \pi \cdot \eta \cdot r \cdot N_A}$$

(3.4)

If a potential (typically 10 to 30 kV) is applied, separation takes place on the basis of different migration velocities of the solutes in the buffer. The ions, accelerated through the electrical field, migrate with a mobility μ that can be calculated from the following equation

$$\mu = \frac{L_{eff}}{t \cdot E} = \frac{L_{eff} \cdot L_{tot}}{t \cdot U} \qquad (3.5)$$

In calculations it is important to distinguish between the total capillary length (L_{tot}) and the length from the inlet to the detection site (L_{eff}), because the electrical field decreases over the total length, whereas the solutes traverse only the effective capillary length during the migration time. Therefore, both capillary lengths should always be given in the form $L:L_{eff}/L_{tot}$.

Table 3-1 Limiting mobilities of some ions important in CE

Cations	Mobilities	Anions	Mobilities
H_3O^+	362,5	OH^-	-205,5
Li^+	40,1	F^-	-57,4
Na^+	51,9	Cl^-	-79,1
K^+	76,0	NO_3^-	-74,1
NH_4^+	72,2	SO_4^{2-}	-82,9
Diethylammonium	37,9	Hepes	-21,8
Triethylammonium	53,1	Mops	-24,4
Ammediol	76,2	Mes	-26,8
Tris	29,5	Amp	-22,6
ß-Alanine	37,5	Formate	-56,6
Ethanolamine	44,3	Acetate	-42,4
Imidazole	52,0	Aces	-31,3
		Mopso	-23,8
		Bes	-24,0
		Tes	-22,4
		Heppso	-22,0

(The mobilities represent limiting mobilities at 25° C in $10^5 \cdot cm^2 \ V^{-1} s^{-1}$).

Tris (Tris-(hydroxymethyl)-aminomethane), Hepes (N-2-Hydroxyethylpiperazine-N'-2-ethansulfonic acid), Mops (3-(N-Morpholino)-propansulfonic acid), Mes (2-(N-Morpholino)-ethansulfonic acid), Amp (Adenosinmonophosphate), Aces (N-(2-Acetamido)-2-aminoethansulfonic acid), Mopso (3-(N-Morpholino)-2-hydroxypropansulfonic acid), Bes (N,N-Bis-(2-hydroxyethyl)-2-aminoethansulfonic acid), Tes (N-Tris-(hydroxymethyl)-methyl-2-aminoethansulfonic acid), Heppso (N-Hydroxyethyl-piperazine-N'-2-hydroxypropansulfonic acid),

Electrophoretic separations are possible only if there is a difference in the mobilities of the ions. The effective charge of a solute ion is the charge of that ion minus the fractional charge of the surrounding oppositely charged ions (in the rigid double layer model). During migration the ion drags this portion of the double layer with it and therefore migrates more slowly than what would correspond to its actual charge. This is called the electrophoretic effect and is largest for thin, diffuse double layers surrounding the ion. This characteristic double layer can be calculated from the Debye-Hückel theory and is inversely proportional to the square root of the electrolyte concentration. It can be shown experimentally that the effective charge and therefore the migration velocity decreases with increasing ionic strength.

As is to be shown later, symmetrical peaks are obtained only if the mobilities of the sample components and of the buffer components are similar. For this reason, the mobilities of the frequently used buffer components are summarized [7] in Table 3-1. As can be seen from it, the buffer components used in classical electrophoresis are marked by low mobilities, whereas inorganic ions exhibit mobilities up to three times greater.

For large particles of similar composition and sizes larger than the double layer, the mobilities are independent of the size, which makes their separation by electrophoresis difficult. Thus, the migration velocity of DNA molecules and of proteins denatured with SDS is nearly identical in pure electrolyte. Separation effects are attained only if the migration is modified by exclusion or sieving effects (e. g., in gels).

3.2 Conductivity

Electrolytic mobility of the ions manifests itself macroscopically as conductance of the current by the electrolyte. According to Kohlrausch's Law of independent ionic migration, all ionic components of the electrolyte contribute to charge transport and hence to the current. The current density J is proportional to the concentrations, charge, and mobilities of the various ionic species. The current density, i. e., the electric current I transported through unit surface area S, is related to the field strength E via Ohm's Law as

$$J = \frac{I}{S} = E \cdot \kappa,$$

(3.6)

where κ is the specific conductance. If the mobilities of the ions are known, the specific conductance of the electrolyte can be calculated from their concentration and Faraday's constant.

$$\kappa = \frac{J}{E} = F \sum_{j=1}^{n} C_j \cdot Z_j \cdot \mu_j$$

(3.7)

In electrochemistry the so-called transference numbers are used, which give the fraction of the current carried by each ion.

$$t_j = \frac{F \cdot C_j \cdot Z_j \cdot \mu_j}{\kappa} \tag{3.8}$$

In the tabulated data of physical chemistry one also finds the molar conductances of electrolytes. These relate the specific conductance to the concentration c of the electrolyte solution.

$$\Lambda_m = \frac{\kappa}{c} \tag{3.9}$$

Analogously, for the molar conductance $\lambda_{m,A}$ of ion A we have

$$\lambda_{m,A} = \frac{\kappa_A}{c_A} \tag{3.10}$$

The usual SI units for molar conductance are S m^{2}mol^{-1}. In practice, units 10^4 larger are frequently used (S cm^{2}mol^{-1}).

The mobility of ions is affected by temperature and ionic strength. The dependence of mobility on temperature is a function almost exclusively of the variation of the viscosity with temperature, i. e., the mobility increases directly with temperature. The change in the solvation shell with temperature contributes but little to the mobility of ions.

The effect of electrolyte concentration on conductivity or ionic mobility is treated in detail in textbooks on physical chemistry. The decrease in the molar conductance as a function of the concentration had been described by Kohlrausch by the empirical equation

$$\lambda_c = \lambda_0 - A\sqrt{c}, \tag{3.11}$$

where λ_o represents the molar conductance at infinite dilution. New types of systems of equations, some of them series such as the Onsager equation, permit more accurate calculations of the concentration dependence but go too far for practical electrophoresis. The literature also contains tables of relative conductances based on one ion, e. g., sodium ion for which the concentration dependence is known or can be calculated via the Onsager equation.

For completeness it should be pointed out that the conductance of weak acids or bases depends on the degree of dissociation of that substance and therefore on the pH of the electrolyte solution. The dependence of the mobility on pH in the vicinity of the pK value of the electrolyte component exhibits a shape similar to that of the titration curve of the component.

3.3 Electroosmotic Flow

Electrophoresis effects the separation of particles of different mobilities, whereas electroosmosis causes the buffer solution to flow in an electric field. In most cases in CE the electroosmotic flow (EOF) is superimposed on the electrophoretic migration of the ions. This EOF depends upon the distribution of charge in the proximity of the capillary surface. Nearly all surfaces carry a charge. In the case of quartz capillaries these are the negative charges from the dissociation of the silanol groups. In solution these surface charges are counterbalanced by oppositely charged ions (counterions). In this double layer, which is presented schematically in Fig.3-1, the positive ions predominate in quartz capillaries and are arranged in a rigid and a diffuse layer. According to Stern's theory [8], the potential built up on account of the charge distribution is divided into two regions as shown schematically in Fig. 3-2: (a) a linear decrease in the potential in the region of the rigid boundary layer, and (b) an exponential decrease in the diffuse boundary layer. This exponential decrease is responsible for the electroosmosis and is designated as the ζ-potential.

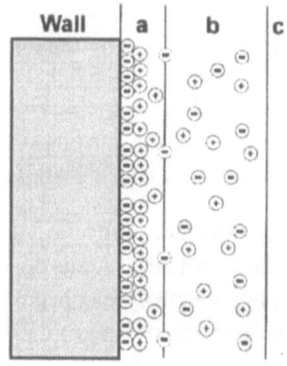

Fig. 3-1
Charge distribution at the surface of fused silica and formation of the ζ-potential

a: rigid boundary layer with adsorbed ions
b: Stern boundary layer (diffuse boundary layer)
c: Electrolyte

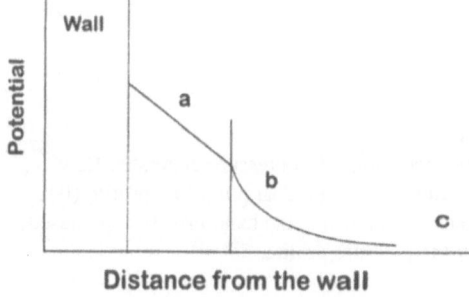

Fig. 3-2
Course of the ζ-potential at the buffer/fused silica interface

a: rigid boundary layer with adsorbed ions
b: Stern boundary layer (diffuse boundary layer)
c: Electrolyte

If an electric field is applied parallel to the surface, the field pulls the counterions of the mobile layer along its axis and thusly moves the entire liquid in the capillary along with it. In quartz capillaries with an enrichment of positive ions in the boundary layer the EOF is induced to move to the cathode. An extremely flat (piston-shaped) flow profile is produced. Calculations have shown that under experimental conditions typical for capillary electrophoresis (10 mM buffer and 30 kV separation potential) liquid layers as little as 10 nm from the quartz surface already move uniformly. This leads to substantially less band broadening than for hydrodynamic flow where the parabolic Hagen-Poiseuille flow profile appears, which is strongly dependent upon the capillary radius and the flow velocity. Fig.3-3 shows a comparison of both flow profiles. In capillaries with inner diameters of 25 μm to 100 μm the flow profile can be regarded as nearly ideally plug-shaped.

In capillaries packed with glass beads or silica particles the EOF should be independent of the particle size of the packing material and have the same direction as in the empty capillary. Therefore, from the theoretical aspect nothing would stand in the way of using very small particles (particle size around 1 μm or less) or longer columns in chromatography. Hence the process of "electrochromatography" is gaining increasing interest because it could combine the selectivity of chromatography with the separation efficiency of CE. By utilizing nonporous particles, the contribution of diffusion in the pores to band broadening can be eliminated.

The migration velocity u of the EOF can be described in simplified form by means of the Helmholtz equation

$$u = \frac{\varepsilon \, E \zeta}{4 \, \pi \, \eta} \tag{3.12}$$

The EOF is inversely proportional to the viscosity of the electrolyte, proportional to its dielectric constant ε, the applied field strength E, and the ζ-potential (zeta-potential). For quartz capillaries the EOF diminishes with increasing electrolyte concentration, with the addition of organic components, and increases with the degree of dissociation of the surface silanol groups, i. e., with the pH value.

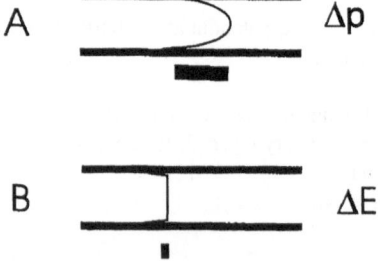

Fig. 3-3
Representation of a pressure-generated flow profile (A) and an ideal plug flow profile (B). Sketched in also is the band broadening caused by the pressure profile.

The electroosmotic flow can be evaluated by injecting an uncharged substance (neutral marker). From the migration time t_m of the neutral marker the velocity and mobility can be calculated from equations 3.1 and 3.5, respectively.

The dependence of the EOF on pH in quartz capillaries and the corresponding reproducibility of the mobility is shown in Fig. 3-4. For a cyclical buffer change the EOF exhibits a typical hysteresis phenomenon. The largest deviations occur in the central pH region in the vicinity of the pK value for silicic acid. The deviations can be reduced, however, by extending the conditioning times following changes in the buffer pH. The hysteresis phenomena are thereby reduced. For reproducible work in untreated capillaries it is therefore essential to standardize the flushing and conditioning times of the capillary to reduce the hysteresis phenomena. It is therefore advisable to carry out some separations after a buffer change before testing the reproducibility in a new buffer.

After conditioning the capillary (3 min with NaOH), the EOF was measured each time after flushing the capillary for 2 min with separation buffers of different pH values to obtain the results presented in Fig. 3-4.

As already explained, the EOF decreases with increasing ionic strength. This is depicted in Fig. 3-5 and 3-6. If the EOF is plotted against the logarithm of the buffer concentration or the ionic strength, a linear relationship is obtained because the ζ-potential is also a function of the square root of the buffer concentration.

The EOF appears in all electrophoretic separation methods because surface charges cannot be completely eliminated. On the one hand, it can lead to convective mixing of the electrophoretic zones, but on the other, it also plays an essential and decisive role in the transport of the zones through the capillary. Because of the ever present EOF, the detector in CE is usually placed on the cathode side. Cations move with the EOF (comigration). Very rapid analysis times are therefore achieved with positively charged compounds. Even anions that migrate opposite to the direction of the EOF (countermigration), are transported to the detector (on the cathode side) if their migration velocity is lower than the velocity of the EOF. Therefore, under suit-

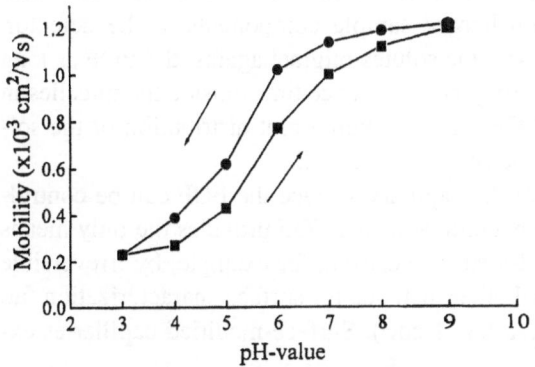

Fig. 3-4
pH dependence of the electroosmotic flow [9]

Separation conditions: capillary: 75 um i.d., L=40/47 cm; buffer: phosphate 10 mM; neutral marker: benzyl alcohol; E = 425 V/cm

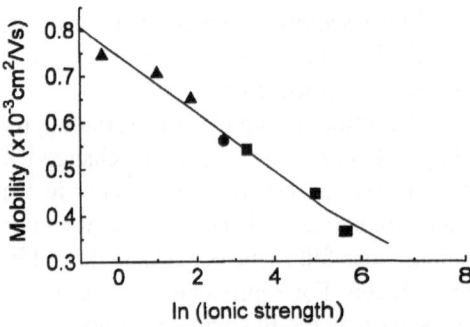

Fig. 3-5
Concentration dependence of the electroosmotic flow. Mobility of the EOF *vs.* the logarithm of the ionic strength of the buffer [10]

Separation conditions: borate buffer (triangles), carbonate buffer (circles), phosphate buffer (squares), each pH 8.0

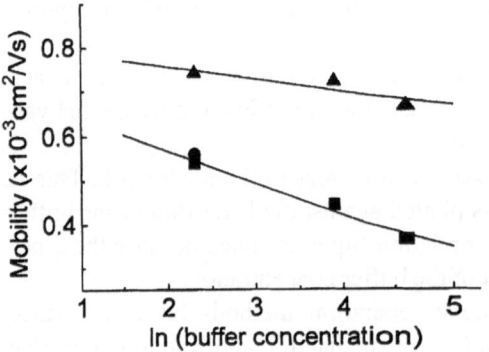

Fig. 3-6
Concentration dependence of the electroosmotic flow. Mobility of the EOF *vs.* the logarithm of the buffer concentration [10]

Separation conditions: borate buffer (triangles), carbonate buffer (circles), phosphate buffer (squares), each pH 8.0

able conditions cations and anions can be separated from each other in a single analysis (cf. Fig. 2-2). Only the anions that migrate faster than the flow velocity of the EOF migrate into the anode compartment and escape detection. These ions can be detected by reversing the polarity, but then the cations and the slowly migrating anions escape detection.

In other separation methods, such as MEKC for example, the EOF is used exclusively to transport the frequently uncharged sample components to the detector. When anionic micelle formation is used, the solutes migrate against the EOF as long as they are inside the micelles, but during their residence time outside the micelles in the electrolyte they move with the EOF. This countercurrent distribution of the solutes is one of the reasons for the high selectivity of MEKC.

Through chemical modification of the capillary surface the EOF can be controlled, nullified, or even reversed. The magnitude of the EOF provides the only means of determining changes at the capillary surface caused, for example, by irreversible adsorption of sample components. All other methods for surface characterization fail because of the very small surface area (ca. 1 cm^2). Surface-modified capillaries exhibit scarcely any hysteresis phenomena after a buffer change. When working with gel-filled capillaries, the EOF should be completely suppressed, for otherwise the gels will be flushed out of the capillary. By modifying the capillary surface, interactions of the sample components with the wall can be suppressed. Sorption effects of sample components at the wall always degrade the efficiency (resistance to mass transfer) and, especially in the separation of proteins, lead to a significant reduction in the plate numbers.

The EOF can be reversed by adding long-chained cationic detergents, such as cetyltrimethylammonium salts that are adsorbed on the surface silanol groups. A double layer of the detergent is formed, with the positive charges directed toward the electrolyte. The formation of a double layer is shown schematically in Fig. 3-7. Reversal of the EOF can also be attained by the addition of polyamines such as spermine. With such coated capillaries, together with reversal of the field, separation of slowly and rapidly migrating inorganic anions can be achieved in a single analysis.

Table 3-2 summarizes the means of influencing the EOF. As described above, the buffer concentration and the pH value represent the most important parameters. Addition of organic components also affects the EOF. Polymers can, in part, be adsorbed so strongly that they are not flushed out during a buffer change. Such cases, as well as the adsorption of surfactants, are termed "dynamic coatings".

The EOF also shows a characteristic pH dependence similar to that in Fig. 3-4 for flow reversal caused by coating with polymers or surfactants. The EOF is largest (now towards the anode) at low pH values. The greater the pH value, the smaller is its strength toward the anode. The additional dissociation of the silanol groups reduces the EOF to the anode. At lower surface coatings with positive charge the EOF can again be in the normal direction toward the cathode, especially at high pH values.

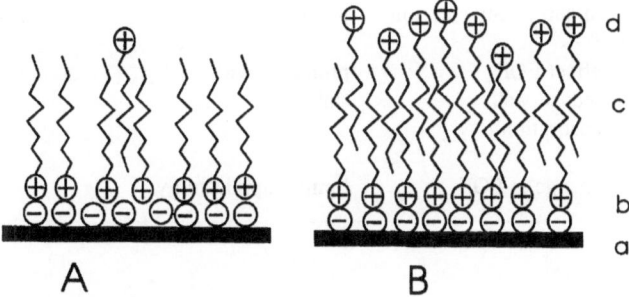

Fig. 3-7 Adsorption of cationic surfactants at the capillary wall: single adsorbed surfactant layer (A), surfactant double layer (B)
(a) capillary wall with negatively charged silanol groups
(b) polar (positive) parts of the surfactant molecule
(c) hydrophobic region
(d) positive surface charge

Table 3-2 Options for modifying the EOF

Change in separation system	Effect on EOF	Remarks
pH-value of buffer	EOF ↗ when pH ↗	may affect the charge of analytes
Buffer concentration	EOF ↗ when buffer conc. ↘	high concentrations cause high currents low concentrations readily lead to overloading
Temperature	changes the viscosity (*ca.* 2-3%/° C)	may affect the selectivity
Organic solvents	change the EOF and buffer viscosity	complex changes in separation system, mostly with additional changes in selectivity
Surfactant as buffer additive	adsorption at capillary wall, drastic changes in EOF	anionic surfactants increase EOF, cationic lower or reverse EOF
Neutral hydrophobic polymers	adsorption at capillary wall, in part strong sorption	reduction of wall adsorption
Ionic polymers	adsorption at capillary wall, in part strong sorption	flow reversal for positively charged polymers
Covalent coatings	affect EOF, reduce wall adsorption	problems with stability
Radial electric field	change in EOF	limited applicability

The EOF can also be influenced by applying a second electric field to the outer wall of the capillary. The capillary wall serves as the dielectric of a condenser. However, this instrument-intensive method for controlling the EOF is effective only at low pH and buffer concentrations. At pH values below 3 the EOF can be reversed in this way.

Variation in the buffer concentration presents one of the simplest and most effective means of influencing the EOF of the separation system. In order to demonstrate the effect of buffer concentration on a separation, a test mixture containing ions of different negative charge that migrate against the EOF, was analyzed at constant current and at constant potential in borate buffers with concentrations between 5 and 100 mmol L^{-1}. The results are presented in Fig. 3-8 and 3-9. This series of measurements shows clearly that the EOF (peak 1 — a neutral substance) increases with decreasing buffer concentration and enables analysis of highly negatively charged solutes that actually migrate counter to the EOF. At constant potential (10 kV) even benzenetricarboxylic acid can still be detected in a 5 mmol L^{-1} buffer, whereas in the same analysis time in a 50 mmol L^{-1} buffer only the singly-charged benzoic acid is transported to the detector by the EOF. Moreover, the current rises from 10 mA to 130 mA. Analogous behavior can be established if these substances are separated at constant current (100 µA). Whereas in a 10 mmol L^{-1} buffer one can operate at 26 kV and all four test substances be conveyed to the detector within 8 min, in a 50 mmol L^{-1} buffer only the neutral marker (benzyl alcohol) could still be detected. All anions migrate faster in the opposite direction. With a pre-set maximum current of 100 µA, one could operate only at 5.5 kV in this buffer. The problems and interferences caused by high currents will be treated elsewhere.

Figure 3-10 shows the course of the migration time as a function of the buffer concentration (constant potential). The highest migration velocities and the shortest analysis times are achieved with the lowest buffer concentration. Since the carboxylic acids migrate against the EOF, their migration time to the detector increases more than that of the neutral marker, benzyl alcohol. To analyze the anions with conventional polarity, one should operate at high pHs and low ionic strengths. If the electrophoretic mobility of benzoic acid is calculated, it is found that it shows no dependence on the buffer concentration. The mobility can therefore be used as a reference value in qualitative analysis when working at different ionic strengths. Fig. 3-10 also shows the increase in current with ionic strength. The linear increase demonstrates the validity of Ohm's Law, which always holds for low heating and good dissipation of Joule heating (cf. 3.4.2).

Furthermore, it has become evident that organic solvents also exert a strong influence on the velocity of the electroosmotic flow. Acetonitrile, methanol, tetrahydrofuran (THF), and isopropanol are presented here as examples. The effects on the EOF were highly variable: 20% (v/v) acetonitrile in the buffer increased the EOF by 15%, whereas 10% (v/v) methanol decreased it by 15% but at the 20% (v/v) level increased it by 5%. THF had little effect on the EOF up to a 20% (v/v) level, and even at 60% (v/v) reduced it by only 20%. Isopropanol produced a drastic reduction in the EOF. At 10% (v/v) the EOF declined to 40% of its original value. No consistency in the behavior of the EOF could be found on the addition of organic solvents, and no correlation with the buffer viscosity could be detected.

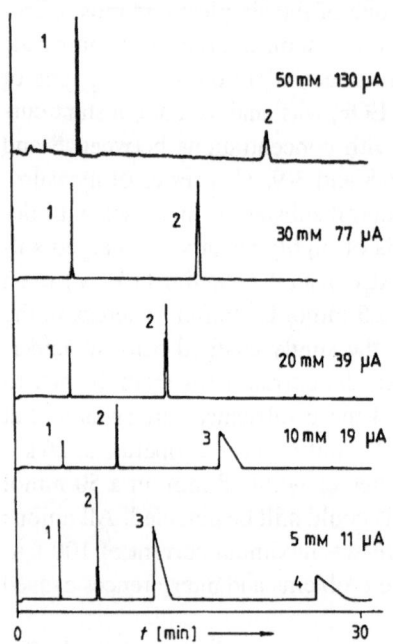

Fig. 3-8
Separation of a test mixture of anions with different buffer concentrations and at constant potential [6]

Separation conditions: Instrument: Beckman P/ACE 2000; Capillary: 75 μm, 37/44 cm; field: 227 V/cm; buffer: borate pH 9.5; injection: pressure, 2 s; detection: 214 nm; sample: benzyl alcohol (1), benzoic acid (2), phthalic acid (3), 1,3,5-benzenetricarboxylic acid (4)

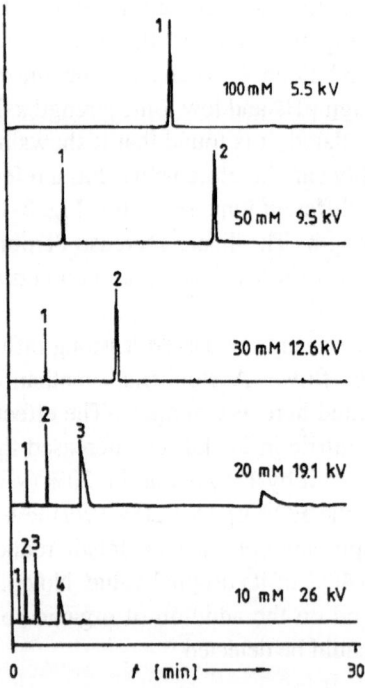

Fig. 3-9
Separation of a test mixture of anions at different buffer concentrations and constant current [6]

Separation conditions as in Fig. 3-8 except field: varied, 100 μA.

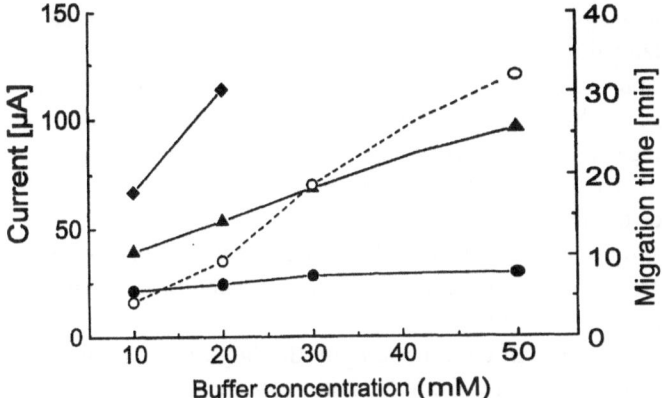

Fig. 3-10 Dependence of the analysis time on the buffer concentration

Separation conditions: migration time of the test anions (solid line for the right axis); current (dashed line for the left axis); benzyl alcohol (circles), benzoic acid (triangles), phthalic acid (diamonds)

3.4 Band Broadening

The well known quantities of chromatography, such as the plate height, H, and the theoretical plate number, N, were adapted for describing zone dispersion in CE. They may be used directly to describe dispersion and transport phenomena in capillaries. Analogous to chromatography, the plate numbers are calculated from the peak width and migration time (eq. 3.13). Since asymmetric peaks occur frequently in CE, the peak width at half-height and the migration time can be used for the practical calculation of the plate numbers according to eq. 3.13:

$$N = 5.54 \cdot \left(\frac{t}{b}\right)^2 \qquad (3.13)$$

(t: migration time of the peak, b: peak width at half-height)

The principal contribution to band broadening in chromatography in open tubes stems from the Hagen-Poiseuille flow profile. This contribution increases with the square of the capillary diameter and is inversely proportional to the diffusion coefficient of the solute in the electrolyte (C-term in the Golay equation) [11]. The flow profile in liquids is hardly equalized by radial diffusion because of the low diffusion coefficients. For this reason capillary liquid chromatography cannot be realized with reasonable capillary diameters (>50 µm). In gas chromatography the diffusion coefficients of the analytes in the gas phase are higher by a factor of 10^4 and the parabolic flow profile is rapidly equalized by radial diffusion due to the concentration profile. Capillary gas chromatography is therefore a highly efficient separation technique. Since the flow profile in CE is plug-shaped as a result of the EOF, its contribution to band broadening can be neglected and in the ideal case only the longitudinal diffusion needs to be considered. The individual contributions to band broadening should not be divided into the three terms of longitudinal diffusion, eddy diffusion, and mass transport as in HPLC, since other causes play a predominant role in CE for poor peak efficiencies. Only the contribution of longitudinal diffusion to band broadening can be transferred from chromatographic theory.

It should be noted, however, that in chromatography the largest contribution to band broadening occurs during the transfer from the stationary phase to the eluent at the end of the column. For on-site detection as, for example, in thin-layer chromatography, the zones are of similar widths so that in the column substantially higher plate numbers are generated than in the detector after elution. The migration velocity of the analytes in the column and in the eluent differ by a factor of $1+k'$. In the column the analytes migrate slowly, after desorption on the way to the detector all migrate with same velocity, that of the eluent. In CE — with on-column detection — the migration velocities through the detection window vary. Slowly migrating analytes stay longer in the detection window, i. e., in integration the area should be standardized with respect to the migration time. For these reasons comparisons of the attainable efficiencies of HPLC and CE are limited.

Another difference between LC and CE should be pointed out here. In LC the plate number is independent of sample size provided one works in the linear region of the sorption isotherm, where retention times are also independent of sample size. Such a region can only be observed in CE when the sample is dissolved in the running buffer. The plate numbers are greatly affected by the operational parameters. When the sample is injected from the buffer solution, the plate numbers decrease with increasing injection time because the initial zone gets longer and volume overloading reduces plate numbers. This effect is less pronounced for electrokinetic injection. When the sample is dissolved in water, sample stacking occurs after injection (*cf.* 4.3.5), i.e., the sample is concentrated into a narrower zone, which results in higher plate numbers. In such cases the plate numbers may even increase with larger injected sample volumes. Hence, in contrast to LC, the plate numbers in CE can be used to compare the separation efficiencies of systems only if all experimental parameters, including the instruments, are kept constant.

3.4.1 Efficiency losses through diffusion

If all the other causes of band broadening are initially neglected, then it is seen that the plate number is directly proportional to the electric field E and inversely proportional to the diffusion coefficient D of the solute in the electrolyte.

A relationship between the important quantities of the plate number N, the field strength E, and the diffusion coefficient D (eq. 3.16) is obtained from the well known defining equation for the plate number (eq. 3.14) and the use of Einstein's Law of Diffusion (eq. 3.15).

$$N = \frac{L^2}{\sigma_L^2} \quad or \quad H = \frac{\sigma_L^2}{L} \tag{3.14}$$

$$\sigma^2 = 2D \cdot t = \frac{2D \cdot L_{eff} \cdot L_{tot}}{\mu \cdot U} \tag{3.15}$$

$$N = \frac{\mu \cdot U}{2D} \tag{3.16}$$

where N: plate number, H: band broadening, D: diffusion coefficient of the solute in the separation buffer, U: potential, μ: mobility.

This equation corresponds to the longitudinal diffusion term of the Golay equation. Since the contribution of the restricted mass transfer in the mobile phase (equalization of the Hagen-Poiseuille flow profile) needs not be considered because of the plug form of the EOF, the plate number increases with decreasing diffusion coefficients of the solute, in contrast to HPLC. From these facts the conclusion is generally derived that CE is particularly suited for the separation of biopolymers because they predominantly possess low diffusion coefficients which decrease with increasing molecular weight. An overview of diffusion coefficients of different sample molecules in aqueous solutions are shown in Table 3-3. With increasing molecular weight, mass transport, i. e., diffusion in a concentration gradient, is retarded and diffusion coefficients consequently decrease with increasing molar mass.

Giddings [12] showed that at room temperature and over a wide range of molar masses the equation for the plate number can be approximately reduced to

$$N \cong 20 \cdot z \cdot U \tag{3.17}$$

where z is the effective charge of the solute in the buffer.

With potentials up to 35,000 volts and effective charges between 1 and 10, up to 10,000,000 theoretical plates per meter are attainable. This number shows that CE is superior to HPLC with respect to possible separation efficiency. The predicted high plate numbers were actually verified for DNA molecules in gel-filled capillaries. Now, DNA molecules represent a special case because with their large number of negative charges they do not interact with the capillary surface and as elongated molecules possess very low diffusion coefficients in the gel matrix. For proteins the diffusion coefficients are larger, so that plate numbers up to 3,000,000 are attainable. The achievable plate number may be a measure of the strength of possible interaction of the protein with the capillary wall.

Table 3-3 Diffusion coefficients and molecular weights of solute molecules in aqueous solutions

Analyte	MW [g mol $^{-1}$]	D [cm^{2}s^{-1}×10^5]
Sodium ions	23	1.25
Ethanol	46	1.08
Valine	117	0.74
Tryptophane	204	0.61
Glucose	180	0.56
Cytochrome C	13,400	0.11
Serum albumin (human)	68,500	0.069
Fibrinogen (human)	340,000	0.019
Tobacco mosaic virus	4,059,000	0.0046

In actual cases, other effects in capillary electrophoresis besides longitudinal diffusion contribute to peak dispersion.

The additional causes of band broadening in CE include
- wall adsorption of solutes,
- disruption of the plug-shaped flow profile by temperature effects,
- superimposition of a pressure flow profile on the electroosmotic flow,
- excessively long sample application zones,
- excessively large sample concentration, and
- mobility differences between buffer and analyte ions.

As in HPLC, the variances of the combination of various effects that contribute most significantly to band broadening are also additive in CE. This results in the reduction of the plate number or increase in the zone width.

$$\sigma^2 = \sigma_{VO}^2 + \sigma_{MO}^2 + \sigma_{LD}^2 + \sigma_{DE}^2 + \sigma_{WA}^2 + \sigma_T^2 + \sigma_{\Delta\mu}^2 \tag{3.18}$$

σ^2: variance of the peak width for Gaussian-shaped peaks. Causes of dispersion:

VO: volume overloading	*MO*: mass overloading
LD: longitudinal diffusion	*DE*: detection
WA: wall adsorption	*T*: temperature effects

$\Delta\mu$: mobility difference between buffer and analyte ions

In the following discussion, some of these points are selected for further scrutiny via practical examples and ways are shown of reducing such effects. Of special interest are the overloading effects, the ionic strength of the buffer, wall adsorption, temperature effects, and mobility differences between buffer and analyte ions.

3.4.2 Efficiency loss through temperature effects

The applied field causes current flow in the capillary. The magnitude of this current depends on the specific conductance of the buffer and the capillary diameter, among other factors. Equation 3.19 describes mathematically the relationship between power and other quantities that are important for the separation.

$$P = U \cdot I = R \cdot I^2 = U^2 \cdot d^2 \cdot \frac{\pi \cdot \kappa}{2L} \tag{3.19}$$

where P: power, d: capillary inner diameter, κ: specific conductance of the buffer.

This equation shows the quadratic dependence of the power on the potential and on the capillary radius. For example, doubling the inner diameter of the capillary requires quartering the potential in order to keep the power and therefore the Joule heating constant. This lengthens the analysis time for large diameter capillaries and simultaneously increases band broadening through diffusion of the analyte. The dissipation of the heat generated by the electric power occurs exclusively across the capillary wall, so that a radial temperature and therefore a viscosity gradient results in the buffer perpendicular to the electrophoretic migration.

Heat is dissipated at different rates through different materials. Whereas water has a relatively high heat resistance of $6.0 \cdot 10^{-3}$ W cm^{-1} K^{-1}, heat is rapidly dissipated through quartz (heat resistance $1.4 \cdot 10^{-2}$ W cm^{-1} K^{-1}) Typical values for the temperature difference between the inner and outer capillary walls lie between 0.3 and 0.7 C [13]. As laborious calculations have demonstrated, a radial temperature gradient with a parabolic shape is formed in the separation buffer. The center of the capillary is heated most strongly and can reach temperatures up to 10°C higher than at the wall. The corresponding temperature gradient is represented schematically in Fig. 3-11.

The radial temperature change creates a viscosity gradient that affects the flow profile. As a result, the solute zones migrate more slowly in regions of high viscosity (capillary wall) than in those of lower viscosity (capillary center). The formation of the temperature gradient depends largely on the capillary dimensions, the buffer conductance, and the cooling of the capillary.

Although cooling the capillary increases the temperature gradient, it is necessary to prevent degassing the electrolyte (and interrupting the current) and local overheating (and denaturing proteins). Viscosity differences between capillary wall and center always lead to migration differences and therefore to higher band broadening and loss of resolution. Losses in efficiency from thermal effects diminish with decreasing capillary diameter and with a reduction in the field strength. Therefore, the high field

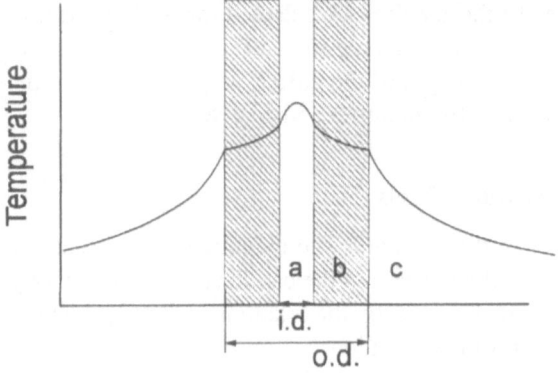

Fig. 3-11
Temperature gradient in the separation buffer and in the capillary wall.

a: parabolic shape in the separation buffer, b and c: exponential temperature decrease in fused silica and in the cooling medium.

strengths needed for rapid analyses can only be applied in narrow capillaries. Consequently, for rapid separations narrow capillaries (50 µm diameter or less) are being used increasingly to avoid interference from this radial temperature gradient. As demonstrated, the temperature difference between the wall and center of a cylindrical capillary increases with the square of its diameter. The temperature gradient itself, however, cannot be measured because of the very small dimensions. The relative loss in efficiency as a function of increasing field strength is displayed in Fig. 3-12 for various capillary inner diameters. As is evident, the smallest losses are to be expected for the 50 µm capillaries. However, with decreasing diameter the thickness of the optical layer and therefore the detection sensitivity (Beer-Lambert Law) are reduced. This, together with the necessary reduction in the sample volume injected, limit the extent to which the capillary diameter can be decreased. The possibility of enlarging the detection cell is being mentioned here already. Another possibility for reducing Joule heating consists of decreasing the buffer concentration and/or using buffers with lower ionic conductances.

In practice, the thermal effects manifest themselves as a reduction in the separation efficiency. As was shown via eq. 3.16, the plate number increases with increasing field strength. This course is shown by curve a in Fig. 3-13. However, the thermal effects also rise with increasing field strength. As a result of the appearance of the zone profiles discussed, band broadening increases and the plate number drops (curve b).

The conductivity of the buffer in the capillary increases with rising temperature and causes the current to change at the beginning of an analysis at constant potential until a steady state and a stable temperature gradient have been established. In this stationary state all of the Joule heat generated is dissipated through the capillary wall. With inefficient cooling the buffer temperature continues to rise and the current increases disproportionately with the applied potential, so that one is operating outside the range of validity of Ohm's Law. In general, the temperature effects on band broadening are negligible as long as one works within the range of validity of Ohm's

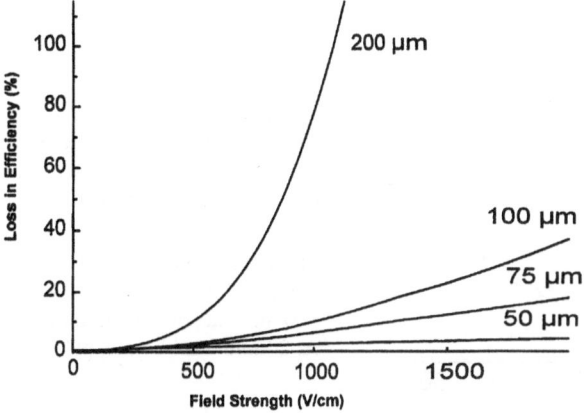

Fig. 3-12
Relative efficiency losses for various capillary diameters as a function of the field strength.

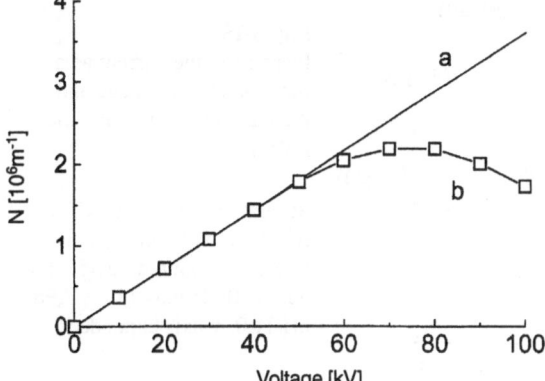

Fig. 3-13
Attainable plate numbers in capillary electrophoresis with or without taking Joule heating into consideration
a: Ideal case taking longitudinal diffusion into account
b: Additional consideration of thermal effects

Law. This can be simply determined experimentally, and the effectiveness of the cooling system of the instrument can be tested in this way. The maximum useable potential and therefore the maximal attainable speed of analysis depends on the efficiency of the cooling system, the inner diameter of the capillary, and the buffer conductance. Fig. 3-14 displays the resulting current-potential curves for a borate buffer for various capillary diameters. Whereas one can operate without difficulty at field strengths up to 20 kV with a 50 μm capillary, for a 100 μm capillary under otherwise identical conditions the Ohm's Law range is already exceeded at field strengths under 10 kV. The advantages of organic buffers with low conductivities are recognizable by comparison with Fig. 3-15. Under identical conditions the 100 μm capillary with CAPS (CAPS = cyclohexylaminopropanesulfonic acid) can be used at field strengths up to 20 kV, whereas with the 50 μm capillary even at 30 kV there is no perceptible deviation from Ohm's Law.

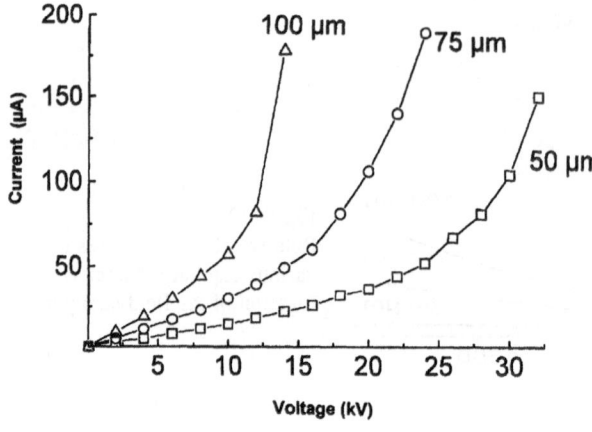

Fig. 3-14
Increase in the current as a function of the voltage and the inner diameter: **inorganic buffer**

Separation conditions: CE instrument: Millipore Quanta 4000; 360 mm o.d., 50/56 cm; buffer: A: 20 mmol L^{-1} borate, pH 10.0.

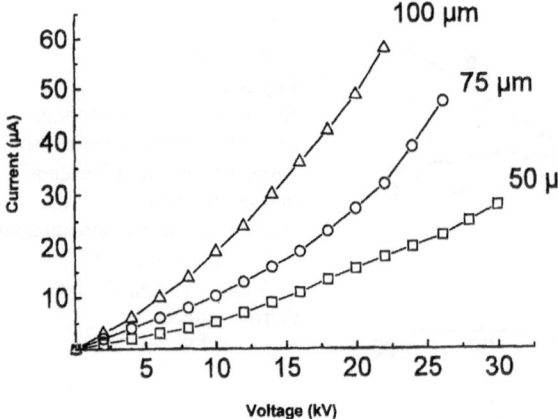

Fig. 3-15
Increase in the current as a function of the voltage and the inner diameter: **organic buffer**

Separation conditions: CE instrument: Millipore Quanta 4000; 360 mm o.d., 50/56 cm; buffer: B: 25 mmol L^{-1} CAPS, pH 11.0; y-axis expanded

3.4.3 Loss in efficiency through electrodispersion

The reduction in conductance of the buffer has its limitations. If large differences exist between the conductance in the buffer and in the solute zones, local interferences with the electric field lead to a distortion of the zones and, in turn, to a reduction in efficiency. If the conductance inside the solute zone is greater than that in the carrier electrolyte, the diminished resistance leads to a reduced field strength. As a result, the solute molecules migrate more slowly at the concentration maximum than at the flanks of the zone. This leads to strong distortion of the zones with a slow rise and a rapid decline (leading or fronting). In the opposite case peaks with strong tailing result. Only when the conductance of the solute zones and the buffer are identical are symmetrical peaks obtained. Therefore the buffer concentration must be adapted to a given separation problem (dissociation and mobility of the analytes).

If the conductance of the solute zone is greater than that of the separation buffer, a dilution of the sample occurs on injection. The reason for this is Kohlrausch's Law which requires a uniform conductance over the entire range of the separation distance, as presented schematically in Fig. 3-16.

If an additional difference in the mobilities between the analyte and buffer ions exists, then additional distortion of the peak shape results from isotachophoretic effects. These complex relationships between the conductance and mobility differences in the analyte zones and the separation electrolyte are clearly summarized in Table 3-4.

In zone electrophoresis the conductance of the buffer should be uniform over the entire separation distance. Only these conditions assure that the potential decreases uniformly over the entire distance and that no jumps in the field strength appear. In

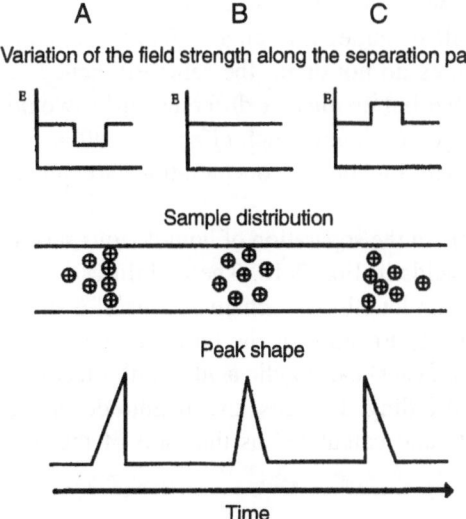

A   B   C

Variation of the field strength along the separation path

Sample distribution

Peak shape

Time

Fig. 3-16
Schematic explanation of band broadening; representation of the variation of the field strength, sample distribution, and the peak shape in the sample zone

A) Sample zones of ions with higher mobility than the background electrolyte
B) Sample zones of ions with the same mobility as the background electrolyte
C) Sample zones of ions with lower mobility than the background electrolyte

Table 3-4 Relationship between peak form and mobilities as well as between the conductances of the sample solution $\lambda_{(s)}$ and the buffer $\lambda_{(b)}$

	$\lambda_s > \lambda_b$	$\lambda_s = \lambda_b$	$\lambda_s < \lambda_b$
$\mu_s < \mu_b$	Dilution, *tailing* peaks	*tailing* peaks	Enrichment, *tailing* peaks
$\mu_s = \mu_b$	Dilution, symmetric peaks	no effect symmetric peaks	Enrichment, symmetric peaks
$\mu_s > \mu_b$	Dilution, *fronting* peaks	*fronting* peaks	Enrichment, *fronting* peaks

the separation of salt-containing samples, such as protein extracts, this causes considerable problems. The ionic strength (or concentration) of the buffer governs the appearance of a field strength gradient in the zone in which the analyte molecules migrate. If the conductance of the analyte zone is not negligible compared to that of the buffer, band broadening results. This effect is amplified by a difference in the mobilities of the analyte and buffer ions. However, as shown above, the Joule heat generated limits the extent to which the buffer concentration can be increased.

Asymmetric peaks due to mobility differences can be improved by changing the buffer concentration, as shown in Fig. 3-17. The H-values of test substances decrease rapidly with increasing buffer concentration. Thus, the H-value for p-hydroxybenzoic acid drops from 31 to 4 μm when the buffer concentration is raised from 30 to 160 mmol. For the inert marker, benzyl alcohol, symmetrical peaks are already obtained at a concentration of 70 mmol. Here only the conductivity differences between the electrolyte and solute zone contribute to peak asymmetry. Even at higher buffer concentrations components with higher mobilities do not attain the same efficiency as solutes with mobilities similar to those of the buffer ions. A different buffer would have to be selected to obtain a similarly high efficiency. Such effects are often observed with indirect detection techniques where low buffer concentrations are generally used.

These problems are particularly noticeable in the separation of homologous series, as exemplified by the aliphatic carboxylic acids in Fig. 3-18. The mobilities of the carboxylic acids decrease with increasing chain lengths. For chain lengths of 6 and 7 carbon atoms the mobilities correspond closely to those of the buffer ions and the peaks are symmetric. Short-chain (and hence faster) carboxylic acids exhibit leading, whereas the long-chain show a pronounced tailing. For this, the magnitude of the asymmetry factor was adopted from HPLC and calculated as the ratio of the dis-

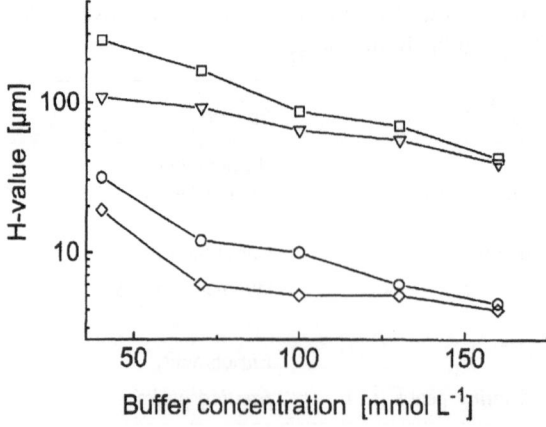

Fig. 3-17
Effect of buffer concentration on band broadening

Separation conditions: CE instrument: Beckman P/ACE 2000, capillary: 50 μm, 54/61 cm; field strength 410 V/cm; buffer: borate, pH 8.5; injection: pressure, 1 s; detection: 214 nm; test ions: phenyltrimethylammonium bromide (squares), phthalic acid (triangles), p-hydroxybenzoic acid (circles), benzyl alcohol (diamonds).

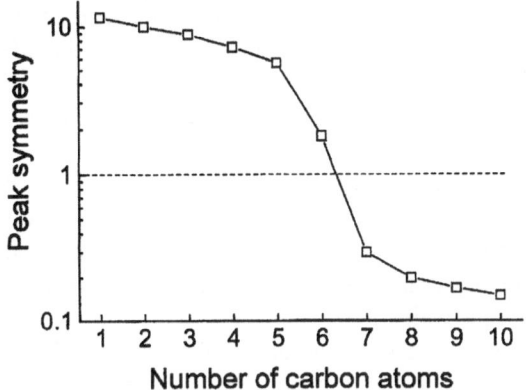

Fig. 3-18
Effect of the mobility differences between analyte and buffer ions as exemplified by the separation of a homologous series of short-chain linear carboxylic acids.

Separation conditions: L=50/57 cm, i.d.: 75 μm, buffer: 5 mM dinitrobenzoic acid, 0.5 mM CTAB, pH 9.0, E=-431 V/cm, detection: indirect UV, 214 nm, sample: homologous series of carboxylic acids, 25 ppm each

tances of the peak flanks from the perpendicular to the peak maximum at 10% of the peak height. Only caprylic acid exhibited a symmetrical peak because its mobility corresponds closely to that of the buffer component dinitrobenzoic acid.

3.4.4 Efficiency losses through wall adsorption

The capillary surface is not inert. Despite the small total surface area, adsorption effects are observed. Each contribution of adsorption leads to additional band broadening and hence to decreased plate numbers. Because of the negatively charged silanol groups of fused silica, positively charged sample molecules are particularly prone to adsorption at the capillary wall. Under neutral and alkaline separation conditions many silanol groups are deprotonated and facilitate adsorption of positive sample ions at the wall. The ζ-potential, which is generated as a consequence of the surface charge on the fused silica, is changed by the adsorption and this results in a variation in the magnitude of the electroosmotic flow which changes the migration times of all peaks. Moreover, the adsorption of sample molecules at the capillary wall reduces peak efficiency and in extreme cases leads to asymmetric peaks with extensive tailing. The evaluation of such peaks is difficult to impossible, and the reproducibility of the analysis is questionable.

Local influences of the surface potential of fused silica may also be expected to change the flow profile from the ideal plug-shaped and cause additional band broadening. This phenomenon can be especially clearly observed for solutes with multiple positive charges. By raising the buffer concentration the ionic interactions between the solute and silanol groups are suppressed, as a result of which symmetrical peaks are obtained.

The suppression of wall adsorption is considerably more important for the separation of proteins by capillary electrophoresis. It could be shown that an increase in the capacity ratio (as a measure of wall adsorption) just from 0.001 to 0.1 leads to an efficiency loss (in terms of increased H-values) from 0.5 to 15 μm [14] because the diffusion of the solute molecules through the diffuse boundary layer (no transport to the detector) becomes the rate-determining step. Practically then, the C-term of the Golay equation must be taken into account.

On repeated injection of protein-containing solutions drifts frequently appear in the migration times. Two fundamentally different strategies are used to solve this problem. In one, a hydrophilic layer is bonded covalently onto the capillary wall, and in the other, substances are added to the buffer that prevent the ionic interactions because they are adsorbed preferentially at the wall.

Figure 3-19 shows the separation of a protein by addition of a buffer additive (1,3-diaminopropane, DAP) that suppresses the interaction between the capillary wall and the protein molecule. The peak of lysozyme becomes increasingly symmetrical with increasing DAP concentration due to its screening effect, and the peak height rises as well.

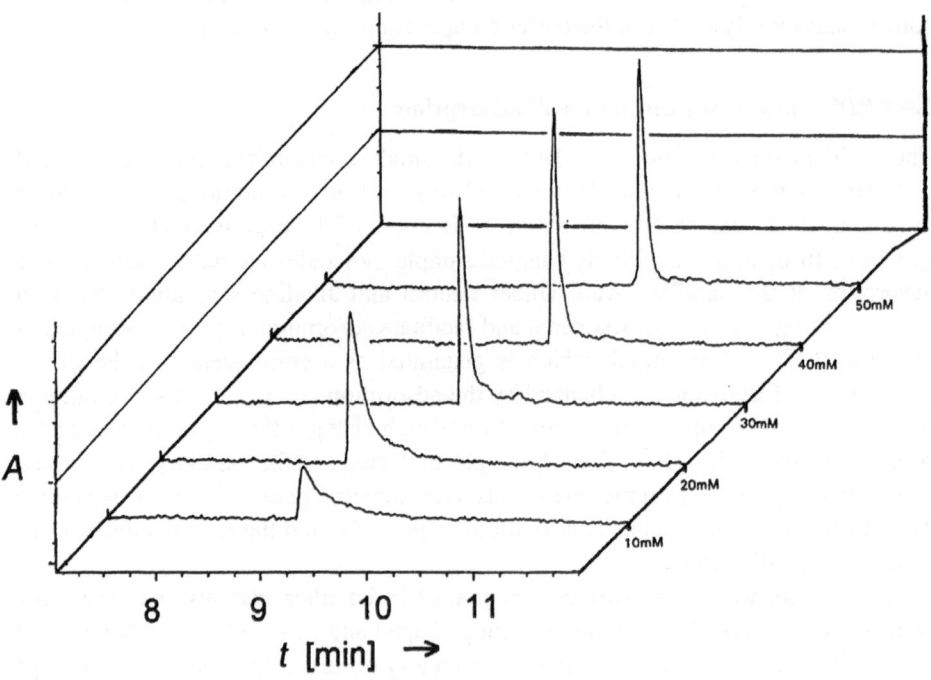

Fig. 3-19. Band broadening due to wall adsorption

Separation conditions: CE instrument: Millipore Quanta 4000; capillary: 50 μm i.d., 42/50 cm; E=300 V cm^{-1}, buffer: 50 mmol L^{-1} phosphate with 20 mmol L^{-1} lithium sulfate and 10-50 mmol L^{-1} diaminopropane, pH 3.5; injection: pressure, 1 s; detection: 214 nm; sample: 0.5 mg mL^{-1} lysozyme

As a result of the reduced wall adsorption, dilution of the substance is progressively diminished and therefore higher and more symmetrical peaks are obtained, even though a relatively large amount of DAP (50 mmol) must be added. Hence, an improvement in the sensitivity can be attained not only through improved detection, but also substantially by lessening band broadening. This example shows clearly that only for high peak efficiency (low dilution) can good detection limits be achieved.

The high concentration of buffer additives cause the conductivity of the buffer to become so large that only low field strengths can be employed for the separation of proteins, which prolongs the analysis times. A detailed discussion of the reduction of wall adsorption follows in Section 5.3 on protein separations.

3.4.5 Efficiency losses through overloading of the separation system

Overloading phenomena are observed when excessive amounts of sample are injected into the separation system. Since no stationary phase is present in CE and separation volumes are limited to a few μL, overloading effects appear readily. Most commonly, improper settings on the instrument or excessive sample concentrations can rapidly lead to overloading. As a rule of thumb, at most 1-2% of the capillary volume may be filled with the sample. For a 50 cm capillary this corresponds maximally to a length of 10 mm and for a 100 μm capillary to a volume of 80 nL. The usual sample volumes lie between 2 and 20 nL.

Beside these volume overloadings for excessively long injection times, mass overloading for high sample concentrations is also observed. Mass overloading is evidenced by plotting the H-values $vs.$ sample concentration. For the same injection time, only the injected amount, not the volume, increases. Volume overloading contributes a constant proportion to band broadening. The relevant curves are shown in Fig. 3-20, where both sample concentration and injection volume were varied.

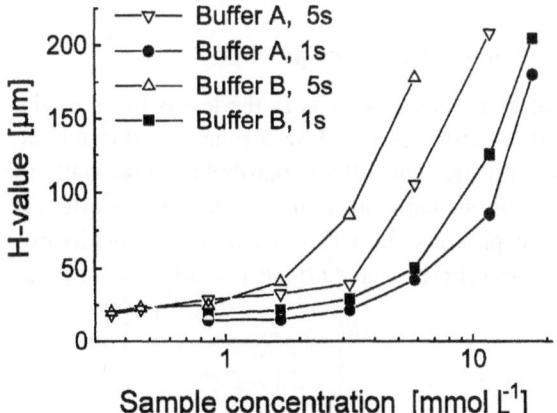

Fig. 3-20
Overloading effects due to excessive sample volume and excessive sample concentration

Separation conditions: CE instrument: Beckman P/ACE 2000; capillary: 75 μm, 65/72 cm; field: 347 V cm^{-1}; buffer: A: 70 mmol L^{-1} borate, pH 8.5; B: 40 mmol L^{-1} borate, pH 8.5; injection: pressure, 1 s or 5 s; sample: phenyltrimethylammonium chloride

Table 3-5. Calculated effect of the length of the injection plug on the band broadening

Length of the injection plug [mm]	N for $D=10^{-5}$ cm^2 s^{-1}	N for $D=10^{-6}$ cm^2 s^{-1}
1	238,000	1,400,000
2	164,000	385,000
10	81,000	112,000

Under the chosen separation conditions of one second injection time, it was possible to operate with sample concentrations of up to about 3 mmol L^{-1}. At higher concentrations the H-value and hence the peak width increases considerably. If the injection time is extended from 1 s to 5 s, the detection limit drops to about 0.2 mmol L^{-1}, but the H-value rises simultaneously because the contribution of volume overloading increases. It is clear that the volume overloading contribution to band broadening is not negligible. Even for low concentrations, in ranges where mass overloading is negligible, the H-values for 5 s injections remain above those for 1 s. For the injection of larger sample volumes, the lower detection limits are compensated by the greater band broadening and the attendant disadvantages in the separation of adjacent peaks. Moreover, it is clearly evident that at higher buffer concentrations the effect of amount of sample and long injection time is less striking (buffer A).

Table 3-5 summarizes the contributions of volume overloading (length of the injected plug) to band broadening. The effect of large sample plugs leads to an intolerable diminution of the plate numbers, especially with high molecular weight solutes. This may pose problems in the analysis of dilute solutions. Possible enrichment techniques (sample stacking) will be discussed later.

3.4.6 Efficiency losses through superimposition of flow profiles

In CE separations care must be taken to avoid differences in the levels between the two electrolyte containers. Even for small differences a flow is generated through the capillary that leads to a parabolic flow profile. This effect contributes additionally to band broadening and depends strongly on the capillary radius. Whereas this effect is generally negligible in 25 μm i. d. capillaries, in 100 μm capillaries the hydrostatically generated flow greatly degrades efficiency and affects resolution and migration times.

3.4.7 Summary

For a better overview, the most important causes of zone broadening in CE are summarized in Table 3-6.

Table 3-6 Overview of the causes of band broadening

Cause of band broadening	Comments
Longitudinal diffusion	Corresponds to the theoretical limit, increases directly with analysis time and the diffusion coefficients, and inversely with the molecular weight.
Thermal effects	lead to convection and to local changes in buffer viscosity
Injection length	should be smaller than the zone generated by diffusion; may be lengthened to improve the detection limit
Wall adsorption of samples	cause of peak tailing and poorly reproducible migration times
Elektrodispersion (mobility difference)	cause of "triangular" peak shapes
Difference in the liquid levels	hydrodynamic flow with the corresponding flow profile

4 INSTRUMENTATION

Commercial instruments have been available since 1988, and the number of distributors is rising [15]. The individual instruments show hardly any essential differences because the actual separation system is very simple. Gradations occur in the area of sample injection, and in the number and type of detectors and the range and user-friendliness of the controlling and evaluating software. No market overview can be presented here, but only the typical requirements of various components of the instruments. The appendix contains a tabulation of the manufacturers and marketers of instruments.

4.1 Power Supply

The potential should be adjustable over the range of -30kV to +30 kV and to maintain the set value as constant as possible. The maximum current should be at least 250 µA, even though substantially higher values are not attained in practice. Moreover, it has proved advantageous to have the choice of holding the potential or current constant, independent of each other. Automatic polarity reversal is necessary for running series of samples by different analytical methods and with differently oriented electrical fields.

A plot of the potential or, preferably, current curve provides indications of possible interferences during analysis and may be helpful for troubleshooting. In commercial instruments, the opening of the analytical compartment automatically shuts off the high voltage source to avoid accidents due to the high potential. For laboratory-built instruments as well as modular commercial units such precautions are indispensable.

4.2 Capillaries

In CE, fused silica capillaries of 50 to 100 µm i.d., drawn from molten SiO_2 are commonly used. In principle, glass or plastic capillaries could be utilized, but these are not always sufficiently transparent in the low-wavelength UV region. Before use, the polyimide coating of fused silica capillaries must be removed mechanically or burnt off at the position of detection. Recently, capillaries with UV-transparent coatings have become available. For most applications untreated and unmodified capillaries are used. The quality of the silica capillaries provided by various producers differ in the exactness and constancy of the inner diameter, as well as the treatment of the inner surface and the UV-transparency in the low-wavelength region. The capillaries also differ in the ratio of the bore to wall diameter. A typical 75 µm capillary with an outer diameter of 370 µm is presented in Fig. 4-1. The outer diameter may vary between 150 and 520 µm. For most separations capillaries with inner diameters between 50 and 100 µm are employed for the reasons already mentioned.

The best bargains are capillaries sold by the meter by various producers. At present, prices are about \$15-20 per meter. Considerably more expensive are single finished capillaries and those already incorporated into a cartridge system. Prices vary here according to the type of capillary (coated/uncoated, with or without a detector cell) and range from about \$30 to \$150, for capillaries in cartridges up to \$250, and with an integrated detection cell (Z-cell) over \$600.

Basically, systems are to be preferred that enable the use of one's own capillaries because otherwise one is limited to capillary lengths, inner diameters, etc. offered by the instrument manufacturer. Newer developments show that capillaries with an integrated detection cell afford advantages. However, the use of these capillaries depends on the instrument available.

In preparing the installation of one's own cut capillaries, it is essential to control the site of the cut. Only straight cuts assure a perfect injection. The commercially available capillary cutters (sapphire blades or ceramic cutters) permit easy treatment of the raw material.

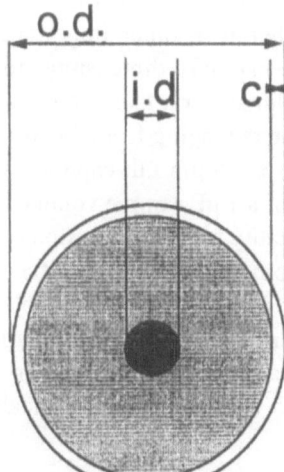

Fig. 4-1
Cross-section of a capillary

i.d.: inner diameter, 10 to 250 µm; o.d.: outer diameter, 150 to 520 µm; c: plastic coating: polyimide or Teflon, 4 to 10 µm thick.

Two options are available for the preparation of the detection window: First, the coating can be burnt off, or second, removed with concentrated sulfuric acid (reaction time about 1 hr). Burning off is performed most easily with a red-hot wire. This method should not be used, however, for capillaries modified with a coating on the inside as this would be damaged as well and make the capillary unusable. Scraping off of the polyimide coating (with a razor blade) is difficult and the risk of breaking the capillary is very large. In some of the new capillary materials UV-transparent Teflon is used in lieu of polyimide to stabilize the capillary. For these capillaries it is unnecessary to introduce a "window" for wavelengths above 200 nm.

The capillaries of various producers differ greatly in the degree of surface hydroxylation that affects the magnitude of the EOF at both high and low pH. The acidity of the surface varies as well, so that the inflection point of a "titration curve" of the base material (cf. Fig. 3-4) lies at different pH values [16]. For new capillaries prior to their first use, it is therefore advisable to standardize the degree of surface hydroxylation by treating them with 1 M NaOH (10 min) and then with 2 N HCl, flushing with water and then equilibrating with the separation buffer for about 20 min.

Surface modification of capillaries can be achieved with the same methods as described for the modification of silica in the preparation of stationary phases for HPLC or the coating of capillary columns for gas chromatography. As already pointed out, for the characterization of modified capillary surfaces, primarily the change in the EOF is available. The characterization of capillaries by gas chromatographic techniques has also been discussed. The advantages and disadvantages of modified capillaries are discussed under the individual separation techniques where they are used. The same goes for gel-filled capillaries.

The typical injection volumes in capillary electrophoresis are between 2 and 20 nL, thereby permitting repeated injections even from samples of only 1 μL. The sample solution after analysis by capillary electrophoresis is available for further investigation. Table 4-1 presents these injection volumes, together with other important characteristics of the separation system that stem from the capillary dimensions, to highlight the problems with small volumes. The injection volumes for a 1 mm long injection plug were calculated for capillaries with inner diameters ranging from 250 μm to 25 μm. The injection volume for the most commonly used 75 μm i.d. capillaries amounts to about 5 nL, which corresponds to 0.05 to 0.2 % of a 1 μL sample volume.

The capillary surface to volume ratio is an important quantity for the discussion of both heat dissipation and chromatographic adsorption phenomena in CE. The surface to volume ratio increases rapidly with decreasing capillary radius. This is favorable for separations because it facilitates heat dissipation and permits higher potentials to be used. It is disadvantageous, however, when samples are adsorbed on the capillary wall. Fig. 4-2 shows the relationship between the surface to volume ratio for typical capillary diameters.

Table 4-1 Characteristic quantities of a capillary

Inner diameter [μm]	Injection volume (1 mm length)	Capillary volume (1 m length)	Relative conductance	Surface per meter	Surface/ volume ratio
250	49.4 nL	49.4 μL	625 %	785 mm^2	11 μm^{-1}
160	20.1 nL	20.1 μL	256 %	502 mm^2	25 μm^{-1}
100	7.9 nL	7.9 μL	100 %	314 mm^2	40 μm^{-1}
75	4.4 nL	4.4 μL	56 %	236 mm^2	54 μm^{-1}
50	2.0 nL	2.0 μL	25 %	157 mm^2	79 μm^{-1}
25	0.5 nL	0.5 μL	6 %	79 mm^2	158 μm^{-1}

In addition to the injection volume, the resistance of the capillary is also an important parameter of the separation system. This value was included in Table 4-1 as a relative conductance value with respect to the capillary cross-section. It is based on the resistance of a 100 μm i.d. capillary. One sees that the relative conductance drops to 25% of its value when the inner diameter is reduced from 100 to 50 μm. This means that under otherwise identical conditions Joule heating is reduced to one fourth. Moreover, since the surface to volume ratio doubles approximately, the heat generated can be dissipated more readily. The use of narrow capillaries has advantages for separations. However, there are also two serious disadvantages: For one, the path length for on-line UV detection and concomitantly the detection sensitivity are reduced; for the other, more time is required to change the separation buffer between analyses.

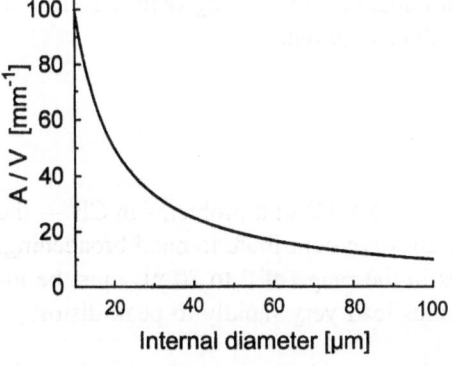

Fig. 4-2
Surface to volume ratio of a separation capillary as a function of the internal diameter

Table 4-2 Time required to exchange the buffer in a capillary as a function of its inner diameter.

Capillary i.d. [µm]	Time required for flushing with a definite volume (ca. 100 µL)	Time required for flushing with 5 times the capillary volume
160	1 min	1 min
100	6 min 33 s	2 min 34 s
75	20 min 43 s	4 min 33 s
50	1 h 44 min 51 s	10 min 14 s
25	1d 3 h 57 min 43 s	40 min 58 s
10	45 d 12 h 16 min	4 h 16 min
5	2 years 4 h 16 min	17 h 4 min

The times hold for a buffer system having the viscosity of water, a 1 m long capillary and a pressure difference of about 0.5 bar.

The reason for the large increase in the rinsing time is the dependence of the flow on the capillary i.d. for a given pressure drop in accordance with the Hagen-Poiseuille Law. The flow (volume per unit time) through a capillary is proportional to the pressure drop and to the fourth power of the capillary radius. Table 4-2 contains the times required to either rinse the capillary with a definite volume (column 2) or with a certain multiple of the capillary volume (column 3).

The times calculated in the table provide an overview of the dependence of the injection times for the introduction of equally long sample zones or for the same sample volumes into capillaries of different i.d. If, for example, it is desired to inject the same sample volume into a 50 µm capillary as into one of 100 µm, a 16-fold longer time is necessary (column 2); if, on the other hand, an equally long sample zone is to be injected, the 4-fold injection time would still be required.

4.3 Sample Introduction

Reproducible sample introduction is one of the most difficult problems in CE — the sample zone should be kept so small that it does not contribute to band broadening. To achieve this, very small sample volumes in the range of 2 to 20 nL must be injected reproducibly. Excessive sample volumes lead very rapidly to peak distortion

Table 4-3 Injection techniques in capillary electrophoresis

	Electro-kinetic injection	Hydrostatic injection	Hydrodynamic injection	Injection via sample-split systems
Injection by:	electric field	siphon-effect	pressure or vacuum	electric sample splitter and split-flow systems
Automation	yes	yes	yes	no
Minimum sample amount	<2.0 μL repeated injection	<2.0 μL repeated injection	<2.0 μL repeated injection	>10 μL via a measuring capillary or HPLC syringe; no repeated injection
Discrimination effects during sample injection	yes	no	no	yes (for electrical splitter) no (for split-flow systems)
RSD values	~3%	<3%	2-3% (from experience)	<3% (ideal case for electric splitter) 2% for split-flow system

RSD: relative standard deviation

and loss in resolution. In order to meet these high demands and to better handle these small volumes, miniaturization and automation of sample introduction is necessary.

The reproducible introduction of small sample volumes is an important aspect of quantification and the standard deviation for analyses. The most important types of injection used in automated commercial instruments are compiled in Table 4-3.

4.3.1 Pressure injection

Samples are introduced by applying a pressure difference between the sample container and the end of the capillary, either by raising the pressure at the sample container or by lowering it at the end of the capillary. The amount of sample introduced, Q, can be calculated from

$$Q = \frac{\Delta p \, \pi \, r^4 \, t_i \, c}{8 \eta L} \tag{4.1}$$

and depends only on the pressure difference Δp and the injection time t_i. For injection times in the seconds range, the pressure difference should be a few mbars. This is the most common injection technique used in commercial instruments because it also allows simple rinsing of the capillary with fresh buffer solution.

Here, the compressibility of gases causes problems. Fig. 4-3 clarifies the problems. For one, the selected injection pressure should be reached quickly, but then the pressure does not drop suddenly at the sample container after injection, so that the exact dosage depends on the volume ratio of the sample to empty space in the sample vial. It is therefore preferable to operate with a pressure/time integral.

The relative standard deviation for this type of sample injection system lies around 2% on the basis of our experience, and can be reduced to under 1% with the use of an internal standard. To determine the injection volume, two principal means are available. One is by calculation, the other is by easy experimental measurement. Calculation of the injection volume is based on the Hagen-Poiseuille Law and is strongly dependent on parameters that are not adequately known. Examples are viscosity and capillary radius. If the capillary radius varies by only 1%, a very large error will be introduced into the calculation of the sample volume because the radius is raised to the fourth power in the Hagen-Poiseuille Law.

The practical determination can be very simply carried out by a breakthrough measurement: The injection time is chosen to be so large that a plug of UV-active solution (e.g., DMSO or benzyl alcohol) is flushed to the detector. The step signal obtained is evaluated by marking the point of the signal at half height and dropping a perpendicular to the time axis, which gives the breakthrough time of the solvent. Since one can work with exactly the same solution as is being injected, errors due to viscosity changes and nonuniform capillary diameters are eliminated. The injection volume is calculated simply from the time required to fill the entire capillary volume and the preset injection time. One would know, for example, from the breakthrough measurement that a flow of 17.7 mm per minute is delivered at a preset pressure. This corresponds to an 8.9 mm long sample plug for a 30 sec injection (typical for ion analysis by CE) or almost 40 nL (for 75 μL i.d.).

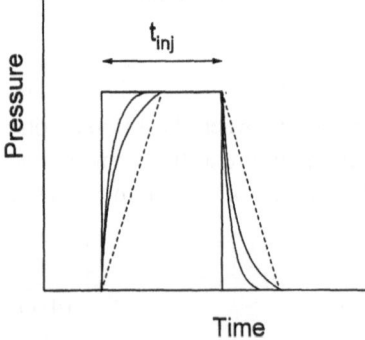

Time

Fig. 4-3
Representation of the pressure/time curves for injection

Rectangular line: idealized curve shape (not realizable)
Curved line: uncontrolled pressure rise, e.g., through simple opening of the pressure valves
Dashed line: controlled rise and fall of a normal pressure/time integral

Table 4-4 Injection volume and initial length of the sample as a function of the capillary diameter and the pressure-time factor.

Pressure-time [mbar·s]	25 µm i.d. Vp [nL]	25 µm i.d. Lp [mm]	50 µm i.d. Vp [nL]	50 µm i.d. Lp [mm]	75 µm i.d. Vp [nL]	75 µm i.d. Lp [mm]	100 µm i.d. Vp [nL]	100 µm i.d. Lp [mm]
25	0,023	0,045	0,375	0,19	1,95	0,44	6,15	0,78
50	0,045	0,090	0,750	0,375	3,90	0,89	12,3	1,56
100	0,090	0,180	1,50	0,750	7,80	1,77	24,6	3,12
150	0,135	0,270	2,25	1,125	11,7	2,66	36,9	4,68
300	0,270	0,540	4,50	2,25	23,4	5,31	73,8	9,36
500	0,450	0,900	7,5	3,75	39,0	8,85	123,0	15,6

Vp: injected sample volume, Lp: length of sample plug
Assumptions: viscosity of water, capillary length of 1 m

This method becomes difficult with very narrow capillaries of less than 50 µm i.d. because the time required to reach the detector becomes very long and a very flat and therefore difficult to evaluate step signal is obtained. The values that are then obtained from the Hagen-Poiseuille Law and the technique presented here sometimes vary up to 30%.

Since various CE instruments permit operation with different injection pressures, it is more straightforward to work with the product of the injection time and injection pressure. An overview of the achievable injection volumes and the lengths of the applied sample zones are presented in Table 4-4.

4.3.2 Hydrostatic injection

In this method a difference in height between the buffer and sample containers is used to introduce small sample volumes. The sample solution is siphoned into the separation capillary. The applied amount of sample depends on the difference in heights (usually 5-10 cm), the time (5 to 45 s), and the hydrodynamic properties (viscosity, density) of the electrolyte solution. Density and viscosity differences between the sample and electrolyte solutions are therefore negligible. The amount of sample introduced can be calculated if instead of the pressure the force of gravity of the solution is taken into consideration:

From $$\Delta p = \rho \cdot g \cdot \Delta h \qquad (4.2)$$

one gets $$Q = \frac{\rho\, g\, \pi\, r^4\, \Delta h\, t_i\, c}{8 \eta L} \qquad (4.3)$$

Accordingly, the applied amount of sample depends only on the height difference and the injection time. Injection times in the range of a few seconds (15-30 s) are appropriate. The values in Table 4-4 can also be used for hydrostatic injection. A height difference of 1 cm corresponds to a pressure drop of about 1 mbar. For a typical injection of 30 s and 10 cm height difference a pressure/time integral of 300 mbar s is obtained, which according to Table 4-4 corresponds to an injection volume of about 23 nL for a 75 μm capillary (total length 1 m).

The height difference should be adjusted accordingly. The calibration curves obtained with this type of injection frequently have a positive y-intercept. This means that even for an injection time of 0 s sample is being injected. Exactly this occurs additionally in some CE instruments during pressure injection: During the time that the sample container is raised and lowered again a portion of the sample flows into the capillary. For excessively short injection times (1 to 5 s) this part of the injection cannot be neglected, as it contributes to the error and may have to be taken into account by calculation. Only for longer injection times (more than about 20 s) can this phenomenon be neglected.

The reproducibility of this sample introduction method is comparable to that of the pressure difference (RSD <3%). For manual sample injection errors of up to 10% are possible.

4.3.3 Electrokinetic injection

In this technique the sample vial in which the capillary is immersed, is connected to the power supply and a potential is applied for a short interval to cause the sample components to migrate into the separation capillary. The amount of sample injected by this technique depends on the magnitude of the applied potential (U_i), the time (t_i) for which the potential is applied, and the mobility of the sample components:

$$Q = \frac{(\mu_p + \mu_{eo})\, \pi\, r^2\, U_i\, t_i\, c}{L} \tag{4.4}$$

where c represents the sample concentration. This relationship underscores the problem with this method, namely the discrimination among the sample components having different mobilities. In extreme cases sample components with $\mu_p - \mu_{eo} < 0$ are not injected.

If the peak area ratios of analytes having different mobilities are compared for electrokinetic and hydrostatic injection, it becomes evident that the faster migrating ion is always enriched in the electrokinetic injection, and the slow ones are discriminated. This is demonstrated for various rapidly migrating cations in Table 4-5, where the area ratios from electrokinetic and hydrostatic injection are compared to each other. The discrimination factor between fast and slowly migrating ions, obtained by dividing the areas in column 2 and 3, correspond exactly to the mobility ratios listed in column 5.

Table 4-5. Comparison of the peak area ratios for electrokinetic and hydrostatic injection [17]

Pair of peaks	Area ratios for electro- kinetic injection	Area ratios for hydrostatic injection	Discrimination ratio	Mobility ratio
Rb^+/K^+	1,0	0,94	1,06	1,04
Rb^+/TMA	2,08	1,33	1,57	1,57
Rb^+/Li^+	1,17	0,69	1,70	1,73
Rb^+/DEA	6,91	3,93	1,76	1,81
Rb^+/Arg	4,34	1,92	2,26	2,31

TMA: trimethylamine, DEA: diethylamine, Arg: arginine

The electrical resistance of the sample solution (ionic strength) in comparison to that of the electrolyte solution also affects the reproducibility of this technique. This can be demonstrated most simply by comparing both types of injection (Fig. 4-4). The difference between hydrostatic and electrokinetic injections is greatest for a sample solution of potassium and lithium ions prepared in pure water (resistance 18 kΩ). The difference decreases with increasing conductance of the sample solution because other ions participate in the charge transfer and therefore fewer sample ions are injected.

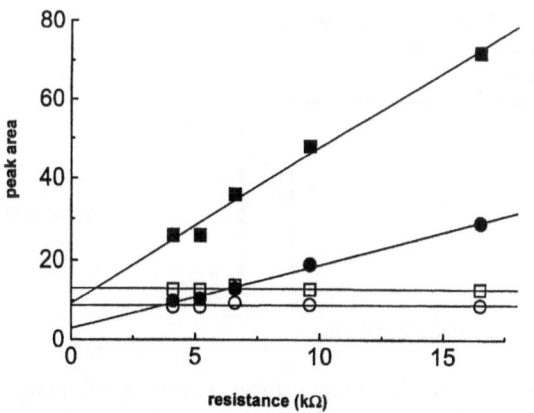

Fig. 4-4
Plot of the peak areas against electrical resistance of the sample solution for hydrostatic and electrokinetic injection [18].

Electrokinetic injection (filled symbols), hydrodynamic injection (open symbols), lithium (circles), potassium (squares).

In the presence of an electroosmotic flow and a low resistance of the sample solution, the ions are injected primarily through the transport of the sample solution (by the EOF), and the electrophoretic migration of the ions then plays only a subordinate role. This figure also clarifies that this effect appears only with ions of very high mobility. The slope of the straight line for the electrokinetic injection of potassium is greater than that for lithium because the former has a greater mobility. In contrast, the slopes of both lines are negligible for hydrodynamic injection: The amount of sample injected is independent of the resistance of the sample solution.

Despite these drawbacks of electrokinetic injection, its use has increased very recently in the form of "electrostacking" that leads to an enrichment of samples by a factor of 10 to 100 and improves the overall detection sensitivity of capillary electrophoresis. Details for the application and optimization of this methodology are given in Section 4.3.5.

A relative standard deviation (RSD) of 4.1% was determined for automated electrokinetic sample application. In general, the reproducibility in quantitative analysis is improved substantially by the use of an internal standard for all techniques that have problems with sample introduction, e.g., capillary gas chromatography.

4.3.4 Sample-split systems

In analogy to gas chromatography, sample-split systems have also been described for CE. Distinction can be made between electrical and hydrodynamic split systems. In an electrical sample splitter, which is presented schematically in Fig. 4-5, there is a branch in the middle of the dosage capillary that leads to the separation capillary. Two different field strengths are applied over both capillaries (dosage and separation). The

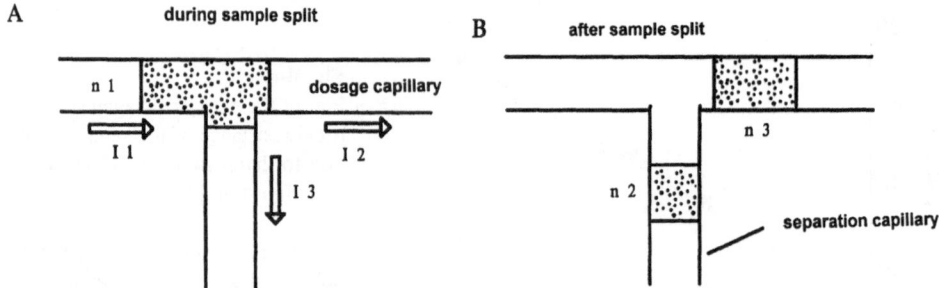

Fig. 4-5 Schematic representation of an electric sample splitter

I_1, I_2, and I_3 correspond to the currents in the respective capillary portions: n_1 corresponds to the amount of sample prior to sample splitting, n_2 to the injected and n_3 to the remaining amounts.

sample moves in two different circuits, with the split ratio being given by the ratio of the two currents. The accuracy of this method was given as <3% [19].

In a flow-split system the sample volume is dosed into a T-joint. The split ratio depends on the ratio of the diameters and lengths of the separation and overflow capillaries. An RSD of *ca.* 2% was stated for this technique, although it requires relatively large sample volumes due to loss during injection and the sample is not available for further testing. Other split systems are known, some using HPLC pumps, but better reproducibility could not be achieved with them either.

4.3.5 Enrichment effects in sample introduction: sample stacking

The problems with the reproducibility of sample injection in CE can, in part, be attributed to the required small pressure differences and short injection times. Under normal conditions larger injected sample volumes rapidly decrease the separation efficiency through volume overloading. One therefore tries to apply larger volumes and simultaneously to sharpen the zone prior to separation. This can be accomplished by utilizing various effects before the actual zone electrophoretic separation.

The sample can be concentrated by working with discontinuous buffer systems. In the simplest case the sample is injected from a purely aqueous solution. As a result of the differences in conductivities between the buffer and the aqueous sample solution, the solutes are first accelerated by a high field strength to the interface between the buffer and the sample solutions, and then migrate more slowly when they enter the buffer region where the field strength is lower. This method is designated as "electrostacking" and is suitable only for salt-free sample solutions.

If no sample solution with high resistance is available, a plug of water can be injected into the capillary before the actual sample injection. The potential drop across the "nonconductor" water is so high that the following sample zone is concentrated in the preceding, substantially higher field.

The potentialities of *electrostacking* are demonstrated [20] in Fig. 4-6. In the upper electropherogram the sample was injected directly from the buffer solution. The analytes become visible only after high amplification. In the middle electropherogram the analytes were injected from a salt-free aqueous solution. The effect of the electrostacking permitted the analytes to be detected after considerable attenuation of the detection signal. A further increase in the peak height can be attained by introducing a short zone of pure water between the electrolyte and the sample (dissolved in the buffer). Enrichment factors over 100 were achieved in this way.

However, in electrostacking the concentration of the sample molecules decreases during the injection process from the inlet of the capillary until an equilibrium state is reached. This causes problems for ions with low mobilities because the permissible length (without contributing to band broadening) of the injection plug is exceeded before the equilibrium state is reached. In these cases the attainable enrichment is lower.

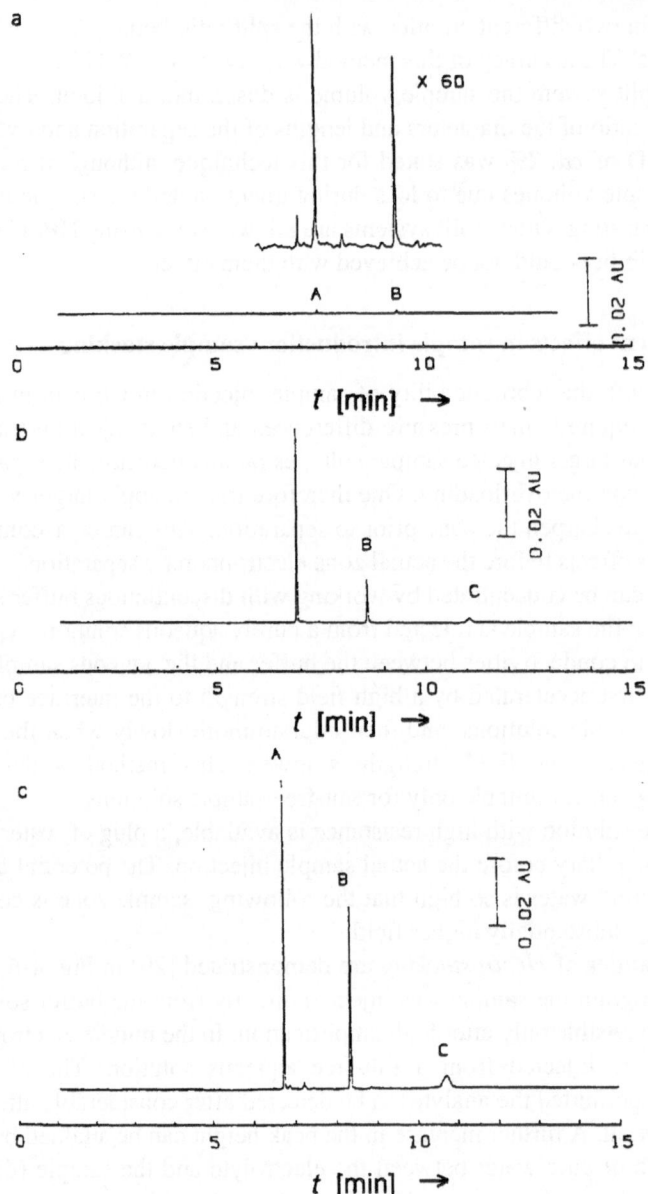

Fig. 4-6 Demonstration of sample enrichment with the separation of PTH-argine and PTH-histidine [20]
a: electrokinetic injection from the buffer
b: electrokinetic injection from water
c: as in b but with an additional water plug between sample and buffer

The EOF also plays an important role in electrostacking. When the EOF and sample migration are oriented in the same direction, the enrichment of the sample is also disturbed. On injection a zone is formed at the inlet of the capillary that possesses very low conductivity. This zone develops more rapidly with increasing magnitude of the EOF and restricts the potential of electrostacking.

If the sample migrates against the EOF, one must distinguish between two cases:

if $\mu_{el} \gg \mu_{EOF}$, good results can be achieved,

if $\mu_{el} > \mu_{EOF}$, the sample cannot be introduced by electrostacking.

In summary, it can be stated that with normal polarity of the CE instrument (outlet grounded) and a cathodic EOF, positively charged analytes can be enriched, whereas in an opposite field anions are enriched on injection.

Samples can be preconcentrated even more effectively if the field is reversed for a short time after hydrodynamic injection. Assuming that the ions to be separated migrate in the direction against the EOF, the capillary can be filled almost up to the detector with the sample solution (hydrodynamic injection) and subsequently removed from the capillary by reversed polarity. The ions migrating against the EOF can be concentrated simultaneously at the interface between the sample solution and the separation buffer. Shortly before this interface reaches the beginning of the capillary, the actual separation can be started by reversing the polarity. The exact time for the reversal of the polarity can be determined experimentally from the current which increases constantly during the preconcentration process. This occurs because the zone of the sample solution (with a high resistance) is transported out of the capillary. When the current reaches about 90% of the maximum value (when the capillary is filled with the buffer alone), the polarity of the potential source can be reversed, and the analyte molecules that gather near the beginning of the capillary, are transported to the detector and separated. Fig. 4-7 shows schematically the separate steps of this injection method, which is generally designated as "stacking with field reversal" [21]. Due to the large injection volume, the sample ions can be concentrated up to a thousand-fold.

Disadvantages of this method are that if the polarity is reversed too late, a part of the sample ions migrate out of the capillary, and that either cation or anions alone can be analyzed.

Other possibilities for preconcentrating the sample consist of using suitable discontinuous buffer systems with different background electrolytes and/or sample solutions with different pH values. These should be designed to change the mobilities of the ions so as to effect preconcentration when they traverse the pH jumps at the buffer-sample solution interfaces. If, for example, peptides are injected from a solution with a high pH and separated in a buffer with a lower pH, the analytes migrate toward the anode because of their negative charge in an alkaline medium, but are protonated at the buffer interface and stopped (due to no effective charge). After the pH steps have disappeared through diffusion and migration of the protons and hydroxide ions, the analytes are separated by zone electrophoresis. This sample enrichment is comparable to that of isoelectric focusing.

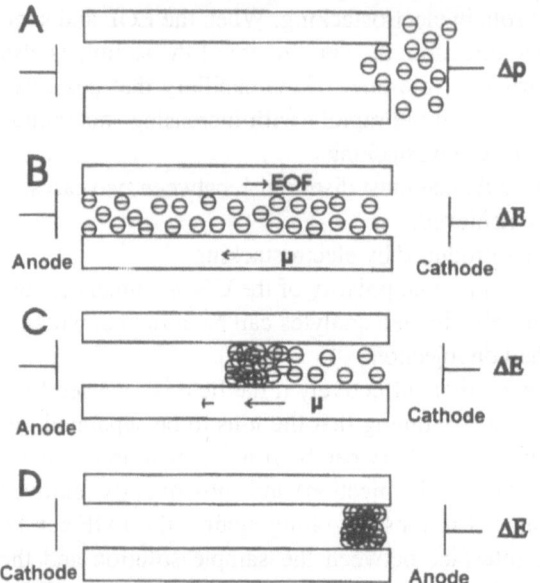

Fig. 4-7
Schematic representation of
sample enrichment with field reversal

A: Pressure injection
B, C: The field is applied to the
capillary filled with sample, and
the anionic components are concentrated
D: Shortly before the sample ex-
its the capillary, the polarity is
reversed, and the analytes migrate toward the detector and are
separated.

Isotachophoresis (ITP) can also be used as an on-line stacking method. The principle of separation itself will be described in a later section. In this stacking process a
leading and a terminating electrolyte must be employed, whose mobilities are slightly
larger (leading electrolyte) and slightly smaller (terminating electrolyte) than those of
the analyte ions. In most cases the separation buffer is used as the leading electrolyte
and after sample injection the capillary is filled with the terminating electrolyte. During the initial time period after applying the potential the sample components, whose
mobilities lie between those of the leading and terminating electrolytes, are sharpened
to a narrow band by isotachophoresis. After a certain time the zone of the terminating
electrolyte is destroyed by the migration and diffusion of the ions, and zone
electrophoretic separation of the ions ensues. This method also yielded enrichment
factors up to about 500.

In this form ITP enrichment of the sample components can be carried out with
commercial instruments. Enrichment is even more effective with two separate capillaries. Initially, the separation is conducted in a relatively wide capillary equipped
with a conductivity detector at the end. The signals generated indicate exactly when
the enriched analytes can be transferred to the separation capillary that is also attached to the end of this capillary. Enrichment factors exceeding 1000 were reported.

Another method is the coupling of a chromatographic enrichment step with subsequent zone electrophoresis. Special capillaries are available for this, whose inlet
side is filled with a few millimeters of a chromatographic phase (mostly C-18). En-

Table 4-6 Overview of sample enrichment methods in CE

Method	Improvement of detection limit	Applicability	Remarks
Electrostacking	10-100	direct	Rapid method development, result depends on the conductance of the sample solution and mobility of sample ions
Field reversal	50-1000	direct	Method development different for cations and anions
Isoelectric focusing	10-100	direct	Useful for weak acids and bases
Isotachophoresis	100-1000	limited	Adjustment of method for each new sample molecule required
Chromatographic enrichment	100-500	only with special capillaries	Expensive and requires capillaries that are difficult to produce

richment occurs when the sample solution is pumped through the capillary and the hydrophilic sample components are adsorbed via the RP mode. These can be desorbed after sample injection with methanol or acetonitrile, which can be transported to the stationary phase by pressure or by the EOF. It is important, however, that the actual separation length between the stationary phase and detector be filled with the running buffer. The advantage of this method certainly lies in its applicability to nonionic, hydrophobic samples. Due to the low reproducibility and high cost of the capillaries, however, this method has not been widely used.

A summary of the methods for improving the detection limits in capillary electrophoresis by on-line concentration of the analytes is contained in Table 4-6, along with some critical comments.

4.4 Thermostating

Thermostating is necessary for dissipating the Joule heating. Air and liquid thermostats are used in commercial instruments, which permit temperature to be varied from about 15°C to about 60°C. In addition to cooling the capillary by ambient air, there are other, more sophisticated means of dissipating the Joule heat. For most applications, however, it is sufficient to remove it by hefty air cooling, where the air flows past the capillary at up to 20 cm s⁻¹. More effective is the removal of the heat with a cooling liquid than with air (heat resistance $2.5 \cdot 10^{-4}$ W cm⁻¹ K⁻¹). Such a cooling liquid is passed over the inlet, detector window, and outlet of the capillary.

Water can be used for separation potentials up to 15 kV, higher potentials require expensive fluorohydrocarbons. What effect temperature has on efficiency and selectivity is still under discussion. One reason for this is that in commercial instruments only the capillary (or parts of it) is thermostated, whereas the buffer containers are not always maintained at the same temperature as the capillary. It was shown for the separation of DNA fragments in gel-filled capillaries that the efficiency decreases with increasing temperature but the relative mobilities, i.e., the selectivity, can be improved. In addition to thermostating the capillary, some instruments are also capable of cooling the sample tray. This is particularly advantageous for the analysis of thermolabile samples. A disadvantage of many such arrangements is that the buffer is also cooled in addition to the sample, which produces a temperature difference between the buffer and the selected temperature of the capillary.

4.5 Detection

In most cases in CE the analytes are detected directly in the separation capillary (on-line), and only in the rarest cases, e.g., with a mass spectrometric, amperometric, or conductometric detector, is the detection performed outside. In contrast to chromatography where the transit time of all analytes through the detector is identical, in CE the sample components migrate at different velocities through the detection window. In the quantitative evaluation of the peak areas (integration of the signal height along the time axis) the different residence times of the solutes in the detection window must be taken into account.

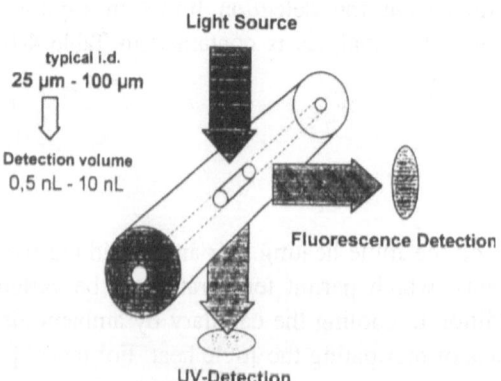

Fig. 4-8
Principle of on-line detection in capillary electrophoresis

Accordingly, only the average width of the capillary inner diameter is available for optical detection, so that the concentration sensitivity is relatively modest. One should not be amazed by published world record detection sensitivities when the detection of single molecules is being discussed. This extremely high mass sensitivity results only from a quirk in calculation when considering the low peak volume of a few nanoliters. Even with the standard concentration sensitivity of optical detectors, atto or yotto moles of a substance can be detected, whereas the concentrations actually lie in the ppb range.

Fig. 4-8 is a schematic representation of the principle of on-line UV and fluorescence detection.

4.5.1 UV detection

Modified HPLC detectors are used most frequently for the measurement of UV absorption. The short path length (the mean capillary diameter) places high demands on the detector with respect to sensitivity, noise, stray light, etc. In order to avoid loss in efficiency due to mixing effects outside the capillary, detection is carried out directly in the capillary. Typical bandwidths of the zone in a capillary are about 5 mm, which corresponds to a volume of 10 nL in a 50 µm capillary.

A more important quantity is the concentration sensitivity. This is rather modest because of the short absorption path (mean capillary i.d.). Compared to HPLC, 30 to 100-fold lower concentration sensitivities have been found for CE. These depend on the detector noise and the effective path length, which deviates downward from the nominal path length (the capillary i.d.). The stray-light portion becomes disturbingly evident through poor focusing (light through the capillary wall) and nonideal cylindrical capillary geometry. These effects can be largely eliminated by optimizing the optics. The very complex light refraction and total reflection encumber the optimization of the optics of UV detectors for CE. Fig. 4-9 shows the calculated light paths for various aperture widths and capillaries and demonstrates the problems involved [22].

Most commercial instruments use either focusing lens systems that irradiate about 0.5 mm or less of the capillary or a beam that has a width of 50 to 200 µm and a length of 100 to 300 µm. The configuration of the capillary in the light path corresponds to that in chromatography. The capillary is irradiated either with monochromatic light or dispersion occurs after the capillary (e.g., for diode array detection).

As in HPLC, fixed or variable wavelength UV detectors are used. Fixed wavelength detectors use a line radiation source whereas deuterium or xenon lamps are employed as continuous sources over the range of 190 to 320 nm. Higher wavelengths are seldom used because very few molecules have high absorption in this range. Fig. 4-10 compares the light intensities of the deuterium lamp and some line radiation sources. The latter produces higher intensity radiation than a deuterium lamp. This illustrates the possibility of optimizing the UV detection by raising the

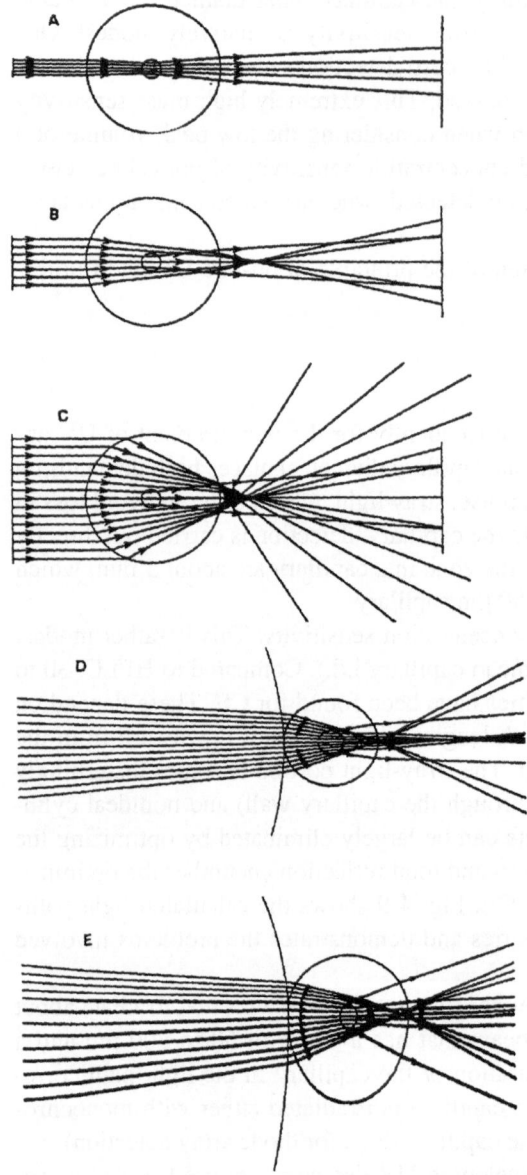

Fig. 4-9 Various detection arrangements for on-line detection [22]

A) Aperture width 50 µm, capillary i.d. 50 µm and o.d. 350 µm
B) aperture width 145 µm, capillary i.d. 50 µm and o.d. 350 µm
C) aperture width 350 µm, capillary i.d. 50 µm and o.d. 350 µm
D) with a lens, capillary i.d. 75 µm and o.d. 275 µm
E) with a lens, capillary i.d. 50 µm and o.d. 350 µm

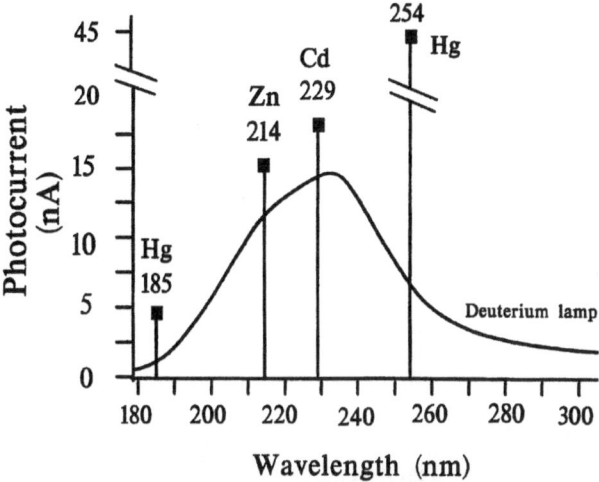

Fig. 4-10 Comparison of the useable light intensities (light intensity measured with a photocell) of a deuterium, mercury vapor, and zinc vapor lamps [23]

amount of incident light. Only in single wavelength detectors are mercury vapor lamps (185 and 254 nm) and zinc lamps (214 nm) used. These lamps can produce about 50 times higher amounts of light than the usual deuterium lamps, and no losses occur through deflection at a grating.

Despite the short optical pathlength, UV spectra can be recorded with sensitive fast-scanning or diode array detectors. The well-known advantages of multi-wavelength detection have been transferred from HPLC to CE. These include:
– facilitation of buffer optimization through automatic peak recognition,
– control of peak homogeneity by spectral comparison within a peak,
– optimization of the sensitivity for substances with greatly different UV spectra by integration of the signals at different wavelengths.

Only a limited number of points per second are measured in the constructive realization of data acquisition with a fast-scanning detector. This causes the detector noise to rise by about a factor of 10 in changing from single wavelength operation to the fast scanning mode for recording spectral information. Moreover, during the appearance of very steep peak flanks (rapid concentration changes) the spectral data could be distorted due to the relatively slow data acquisition.

In contrast, the diode array method corresponds to a real multitasking method with respect to data acquisition. The UV spectra are not distorted by slow data acquisition. Since in CE high peak efficiency and short analysis times lead to steep peak flanks and peak widths of only a few seconds, rapid recording of the spectra is especially important.

1. „bubble"-cell

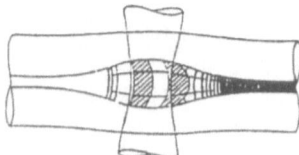

2. „Z"-cell

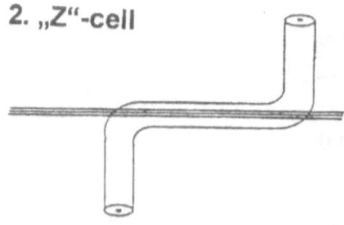

3. rectangular capillaries

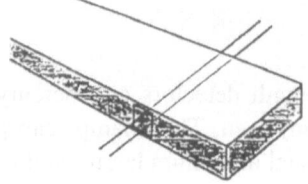

Fig. 4-11
Improvement of UV detection by enlarge-
ment of the path length

1. Bubble-shaped expanded capillary (bubble
cell)
2. Z-cell
3. Capillaries with a rectangular geometry

Many attempts have been made to improve the UV detection sensitivity in capillary electrophoresis by increasing the detection cell path length. Various experimental
approaches have been described that include rectangular capillaries and Z-cells from
micro HPLC. In addition, widening of the capillary at the detection site enhances the
detection sensitivity. The various possibilities for improving detection sensitivities
are shown schematically in Fig. 4-11.

The use of rectangular capillaries improved the detection sensitivity up to 10
times. Practical application of this approach could not be achieved, however, due to
major problems during the injection and bubble formation in the corners. In contrast,
the use of the simple Z-cell for CE failed to improve the detection sensitivity because
the increased signal was accompanied by greater noise from stray radiation. The
newest developments with this system show that the stray light can be minimized
with a spherical lens on the side of the light source directly in front of the bend in the
capillary [24]. An 11-fold enhancement in the detection sensitivity has been reported.
However, there is a problem with this type of selectivity improvement, especially
when high efficiencies are obtained and the peak resolution is small. It can be calcu-

lated that when a peak with 500,000 plates passes through the detector cell with a migration time of 5 minutes, it takes up only 1 mm in the capillary. The resolution achieved with peaks closely following each other can be degraded by this detection technique.

The gain in detection sensitivity and the possible loss of peak resolution is demonstrated in Fig. 4-12, where the separation of pharmaceuticals is compared in a 50 μm i.d. capillary (A) and a 75 μm i.d. capillary with a 3 mm Z-cell (B). In the upper pherogram the height of the last peak corresponds to approximately 1.5 mAU, whereas in the lower its value is 8 mAU.

The gain in sensitivity is much higher because the noise level in the two separations is different. On the other hand, the loss in resolution can be observed by comparing the distances of peaks 1 and 2, and that of the two largest peaks. The rule derived from HPLC that the peak volume should be 5 times the cell volume (because of flushing problems) can certainly not be applied to CE because that would mean a 1.5 cm long zone for the 3 mm cell. But one should always bear in mind when using the Z-cell that closely-spaced adjacent peaks of high efficiency may be missed.

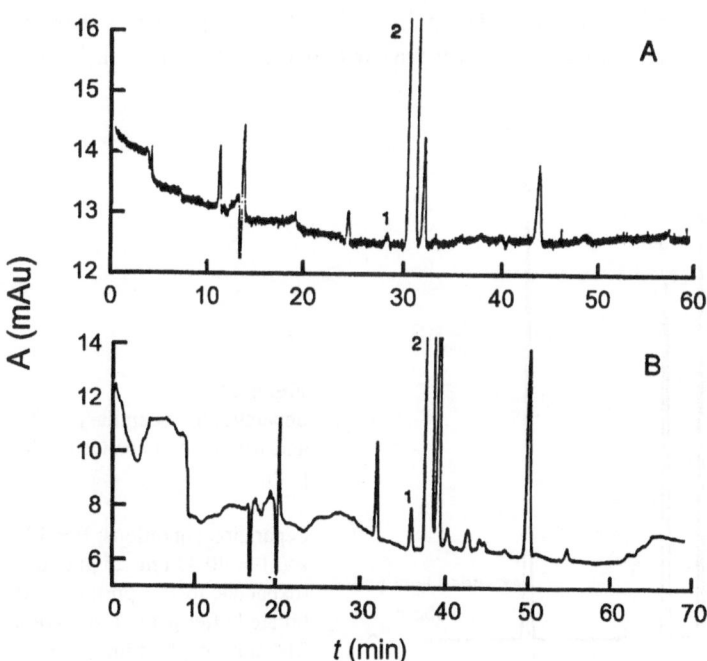

Fig. 4-12 Comparison of detection sensitivity in different capillaries [24a]
A: 50μm i.d.   B: 75μm Z-cell capillary
Length: 72 cm. UV at 220 nm
Buffer: 20 mM Tetraborate pH 9.0; 50 mM SDS; 25% THF
Conditions: 20 kV in A; 15 kV in B

This problem is less relevant for capillaries with bubble-shaped detector cells [25] as the peak volume remains approximately constant while traversing the detector cell but the peak length becomes shorter. With increasing inner diameter, the peak length should automatically become shorter. This holds not only for the flow of analyte zones through a detector cell, but also for the electrophoretic migration of analytes as these move along the field lines through the entire volume of the detection cell.

These devices are commercially available and, in part, permanently attached to the capillary. The price of such capillaries is correspondingly high. An expanded light path can also be produced by simply etching with hydrofluoric acid. A small continuous flow of hydrofluoric acid is generated through the capillary, which is cooled in an ice bath to reduce the reaction rate. At the site where the detection window is to be widened, the capillary is heated with water vapor to about 100°C. In about 10 min. a 3- to 4-fold expanded capillary diameter is obtained at this site. Clearly, this method is applicable only to uncoated capillaries. Since the wall strength of fused capillaries is relatively high, diameters may be safely widened by a factor of 4. By using narrow-bore capillaries (25 μm) good detection sensitivity is obtained with such expanded detection windows. The gain in detection sensitivity is demonstrated in Fig. 4-13, where a 25 μm capillary was widened by etching with hydro-

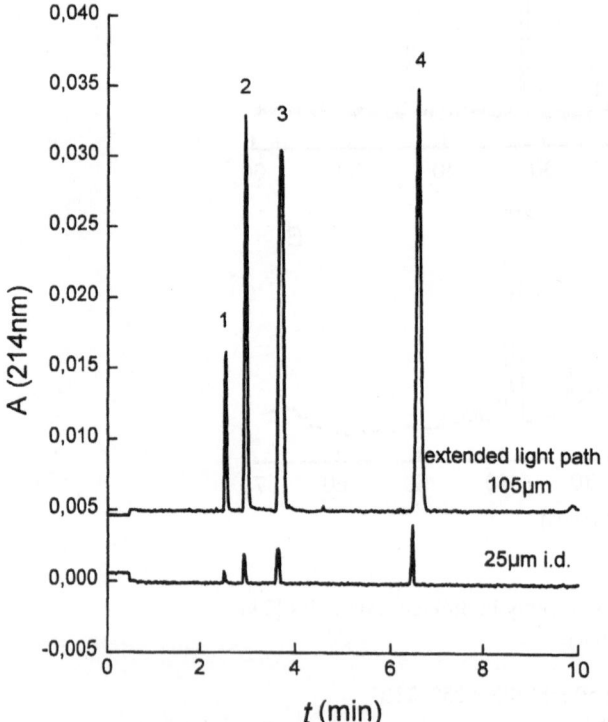

Fig. 4-13
Separation in a capillary with and without an expanded light path

Separation conditions: E = 13 kV; L = 40/47 cm; 25 μm i.d. (expanded to 105 μm); 40 mM borate buffer pH 9.4, detection 214 nm: sample: trimethylammonium bromide (1), p-aminopyridine (2), benzyl alcohol (3), benzoic acid (4).

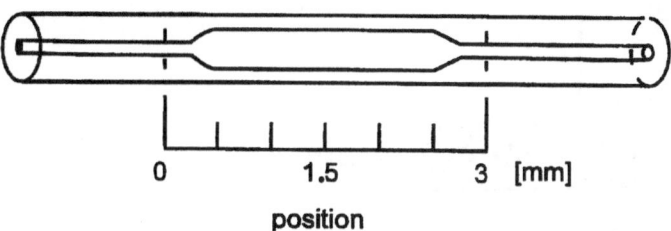

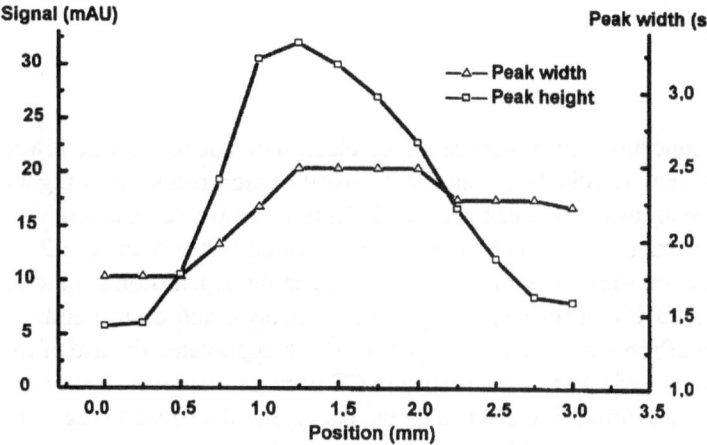

Fig. 4-14 Dependence of the signal height and the peak width on the detection position in a capillary with an expanded light path

fluoric acid to a detection path of 105 μm. This improvement is in accordance with the Lambert-Beer Law.

Since additional band broadening arises in HPLC after each change in the column cross-section, there has been similar discussion in CE. As Fig. 4-14 shows, the gain in detection sensitivity outweighs the band broadening, especially if the measurement is made at the beginning of the expanded site. The signal reaches its maximum value about 1 mm after the start of the expansion and decreases slowly thereafter since dilution of the sample due to the additional band broadening becomes appreciable. Whereas the signal increases by a factor of 6, the peak width (at half height) increases only from 1.2 to 3.6 s.

Analytes that do not absorb in the UV region can be sensed by indirect UV detection with commercial instruments. To effect this, a UV-absorbing electrolyte with a mobility similar to that of the analytes is added to the buffer. Because of the requisite

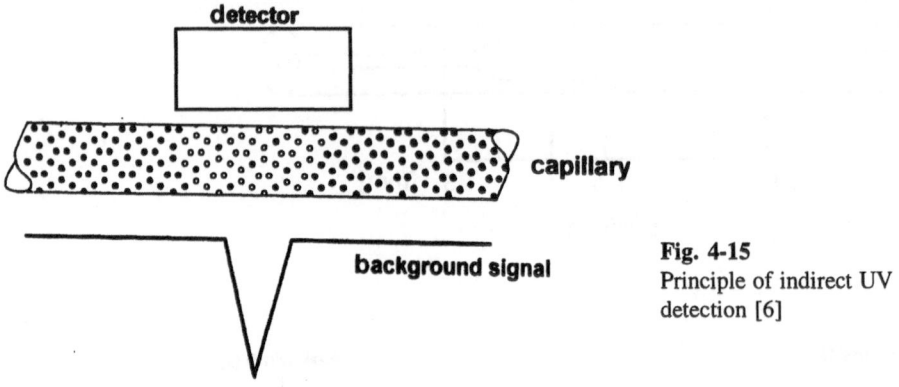

Fig. 4-15
Principle of indirect UV
detection [6]

electroneutrality, the concentration of the absorbing electrolyte added must be lower at the position of the analyte (displacement mechanism), which results in a higher transmittance and is manifested as a negative peak. This is shown schematically in Fig. 4-15. Details and examples of applications are presented in the Section 5.2 on ion analysis. The detection sensitivity of indirect UV detection depends on the molar absorptivity of the added UV-absorbing background electrolyte and corresponds to normal UV absorption. Problems with system peaks, which aggravated the use of indirect detection techniques in HPLC, do not arise in CE because there is no stationary phase whose sorption equilibrium is disturbed by the passage of the solvent plug from the sample. In CE this zone moves with the speed of the EOF, whereas ionic analytes migrate either ahead of or behind the EOF.

4.5.2 Fluorescence detection

In addition to UV detectors, fluorescence detectors are also commercially available. The first approach was the adaptation of HPLC detectors to CE requirements by accommodating the optics to the small volumes. Fluorescence detectors specially developed for CE exhibit substantial differences in the area of light sources. Besides the usual deuterium and xenon arc lamps, laser systems are offered with excitation in the visible wavelength region. These laser fluorescence systems are used primarily for DNA sequencing where the analytes already carry the corresponding dye for detection. Lasers for excitation in the UV region are still expensive and have relatively short lifetimes. Of course, special derivatizations have been described that permit the use of laser induced fluorescence (LIF) detection for small molecules as well.

The gain in sensitivity possible between UV and LIF is demonstrated in Fig. 4-16 via the analysis of oligosaccharides derivatized with 8-aminonaphthalene-1,3,6-trisulfonic acid (ANTS) (reductive amination) [25a]. An improvement in the detection sensitivity by a factor of 300 could easily be obtained.

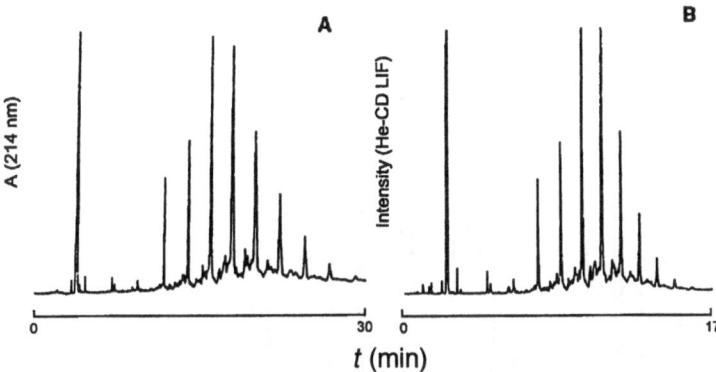

Fig. 4-16 Analysis of malto-oligosaccharides as ANTS derivatives by UV and LIF detection
Conditions: A: UV detection at 214 nm; capillary 670 mm; 20 μm i.d.; 200 mM phosphate buffer
pH 2.0; 30 kV; **20 ng** of mixture injected
B as A except capillary 470 mm; 15 kV; He-Cd laser induced fluorescence; emission at 525 nm;
70 pg of mixture injected

Indirect fluorescence detection is also possible, although some additional problems arise with it. The concentration of the fluorophore cannot be too high because
quenching effects disturb the detection sensitivity. The useful concentration range for
samples is therefore relatively limited. The conventional fluorophores are frequently
relatively large molecules whose mobilities are correspondingly low. Detection sensitivities 20 - 50 times less than those of indirect UV can be obtained even without laser excitation. Aminoacridine, salicylic acid, and Ce(III) compounds are suitable for
the indirect fluorescence detection of small molecules.

4.5.3 Conductivity detection

Because the detection cells for CE must necessarily be small, conductivity and other
electrochemical detectors offer great advantages. A striking example is the detection
of traces (down to 10^{-9} mol L^{-1}) of alkali and alkaline earth metals with
microelectrodes directly in the capillary [26]. Ewing succeeded in directly detecting
neurotransmitters in single cells by amperometry, using 5 μm i.d. capillaries for the
separation [27]. Some approaches for conductivity and amperometric detection are
becoming available now, but whether they find their way from research laboratories
into routine work cannot be forseen at this stage. Amperometric detection would become a very sensitive technique for easily oxidizable or reducible solutes. Some technical problems were encountered with the exact placement of the electrodes within the
capillary, but these seem to have been solved now. The main problem in conductivity

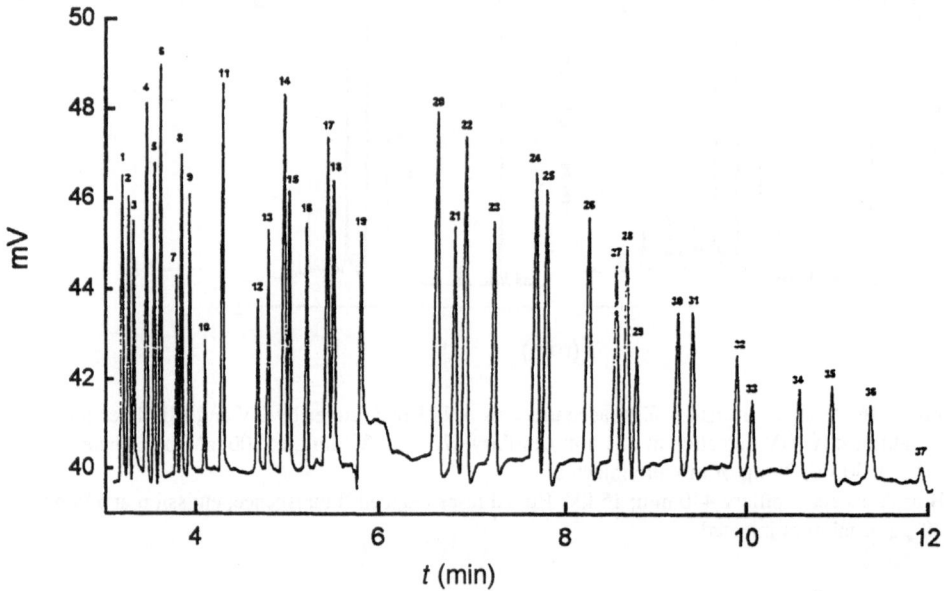

Fig. 4-17 Conductivity detection of 37 organic and inorganic anions
Separation conditions: potential: 25 kV, 10 µA, buffer: 50 mM CHES, 20 mM LiOH monohy-
drate, 0.03 wt % Triton X-100, pH 9.2, 1 mM CTAB, detector: Crystal 1000 conductivity detector (ATI-UNICAM), sample: 4 ppm bromide (1), 2 ppm chloride (2), 7 ppm ferrocyanide (3), 4
ppm nitrite (4), 4 ppm nitrate (5), 4 ppm sulfate (6), 2 ppm azide (7), 3 ppm oxalate (8), 5 ppm
molybdate (9), 6 ppm tungstate (10), 7 ppm 1,2,4,5-BTC (11), 1 ppm fluoride (12), 5 ppm tar-
trate (13), 10 ppm selenite (14), 4 ppm phosphate (15), 5 ppm citraconate (16), 10 ppm glutarate
(17), 10 ppm phthalate (18), 4 ppm carbonate (19), 10 ppm acetate (20), 10 ppm chloroacetate
(21), 20 ppm ethanesulfonate (22), 15 ppm dichloroacetate (23), 15 ppm propionate (24), 20
ppm propanesulfonate (25), 15 ppm crotonate (26), 20 ppm butanesulfonate (27), 15 ppm butyrate (28), 15 ppm toluenesulfonate (29), 20 ppm pentanesulfonate (30), 15 ppm valerate (31),
20 ppm hexanesulfonate (32), 15 ppm caproate (33), 20 ppm heptanesulfonate (34), 35 ppm
MES (35), 20 ppm octanesulfonate (36), 40 ppm d-gluconate (37).

detection involves the determination of the minutely higher conductance of the
analyte zone above that of the background electrolyte. Suppression techniques as
used in HPLC are not applicable here. However, by appropriate selection of the
buffer components with low intrinsic conductances, ions in the lower ppm range can
be detected. The separation of an artificial mixture of organic and inorganic ions is
shown in Fig. 4-17. The highest detection sensitivity is achieved with the ions that
exhibit the highest conductance. The more slowly an ion migrates, the smaller becomes the conductivity difference between the sample zone and the background electrolyte, and the larger is the amount of sample that has to be injected (compare, e.g.,

4 ppm for bromide (peak 1) and 40 ppm for gluconate (peak 37)). Indirect conductivity detection (high buffer conductivity and low analyte conductivity) has thus far not been applied in CE.

4.5.4 Other detection methods

Coupling of CE with mass spectrometry appears to pose few problems when the low flow rates (100 nL min⁻¹) are considered [28]. The main problem in the coupling is to prevent the eluent from being sucked out of the capillary at the transition to the ion source by the prevailing vacuum. For a pressure drop of one bar, a linear flow velocity of 1 cm s⁻¹ is generated in a 1 m capillary (50 µm i.d.). The resulting parabolic flow profile would lead to a noticeable loss in efficiency. For this reason, a make-up flow must be passed around the capillary, which also serves to attain the flow required for electrospray ionization that is about 50 times higher than the EOF in CE. Also, gas must be added for the spray. This shows that the construction of such devices is complicated. By the use of electrospray ionization, biopolymers can be detected through multiple charges. Presently, the most exciting developments in instrumentation are occurring in the area of the direct coupling of CE with MS, favored by novel ionization techniques in MS and by the great need for additional structural information on the separated components, especially in the area of peptides and proteins.

Other detection methods (Raman, on-line radioactivity measurements, circular dichroism, refractive index, and capillary vibration) have been introduced. Their potentials for routine applications cannot be assessed at this time, but a significant improvement in the detection sensitivity is not to be expected.

4.5.5 Derivatization reactions

Chromophores or fluorophores can be introduced into UV-inactive and nonfluorescent analytes prior to separation with the well known derivatization reagents. However, all the disadvantages of pre-column derivatization known from HPLC appear. These include primarily the dependence of the reaction rate on the matrix and problems with polyfunctional sample molecules for which different derivatives could result. Special reagents for electrophoresis that contain charges in addition to the chromophore for a successful separation are being discussed. The most important derivatizing reagents are compiled in Fig. 4-18.

For laser fluorescence detection, the analyte must be reacted in most cases with a fluorescence label having an excitation wavelength above 380 nm. As an example, the separation of the 3-(4-carboxybenzoyl)-2-quinoline carboxyaldehyde derivatives of amino acids can be mentioned [29].

With CE equipment and the possibilities offered by the various injection techniques, it is easy to perform sample derivatization directly in the capillary before a

a) Dansylchloride:

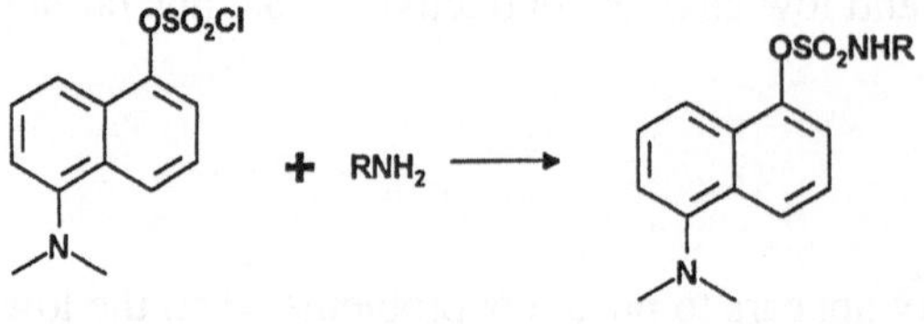

b) FMOC:

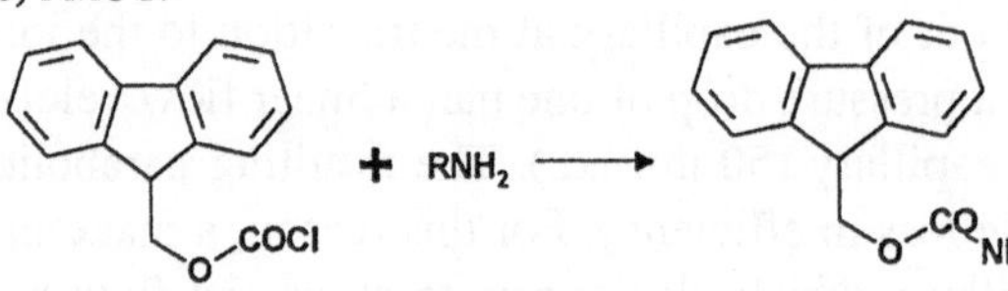

c) Fluorescamine:

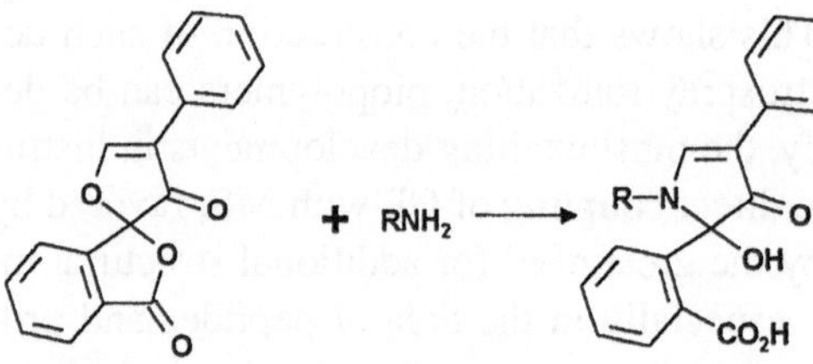

d) CBQCA

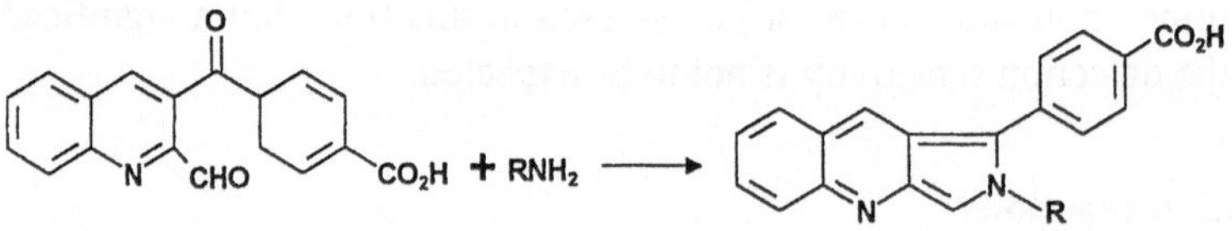

e) OPA

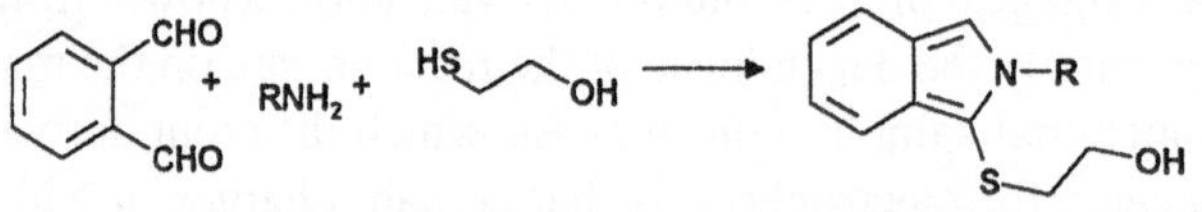

f) reductive amination using p-aminobenzoicacid derivatives

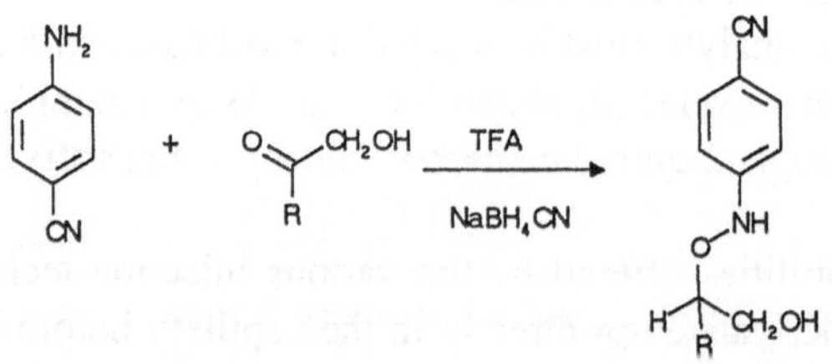

Fig. 4-18
Compilation of some
useful derivatization
reactions

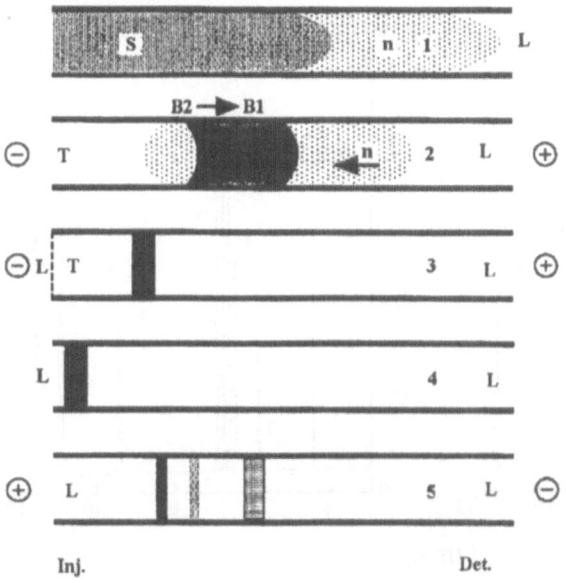

Fig. 4-19
Schematic representation of
on-column derivatization in
CE (for explanation see text)

Inj.   Det.

separation is initiated. A possible arrangement is shown schematically in Fig. 4-19
[29a]. As ususal, the capillary is filled with a leading electrolyte (2). The neutral
derivatization reagent (n) is injected hydrostatically. Then the sample (5) to be
derivatized is also injected after the reagent (1). The capillary is transfered to a vial
containing a terminating electrolyte. After application of the voltage (2), the sample
migrates through the reagent plug where the derivatization reaction takes place.
When leading and terminating electrolytes have been properly selected according to
the rules of ITP, the trailing end of the sample zone (B2) is concentrated at the opposite end of the zone (B1), resulting in a sharp zone of derivatized sample. During this
concentration step the terminating electrolyte and the derivatized sample migrate into
the capillary. The first part of the capillary is filled with terminating electrolyte (3).
This has to be removed by applying presure to the vial filled with the leading electrolyte until the zone of the derivatized sample is pushed back to the front of the capillary. Then the capillary is transfered into the vial with the leading electrolyte (4) and
the system is changed from ITP to CZE and separation can be achieved (5).

The use of isotachophoretic effects helps to sharpen the initially wide solute zone
in the reagent so that a sharp starting zone is achieved and thus the peak broadening
contributions of hydrostatic introduction of reagent and sample zones are eliminated.
When sample concentrations are high, the ITP steps can be readily omitted and the reaction performed in the capillary immediately after sample introduction. As can be
seen in Fig. 4-20, similar peak heights are obtained whether the derivatization of
amino acids with OPA (see Fig. 4-18) is conducted off-line (A) or on-line in the capil-

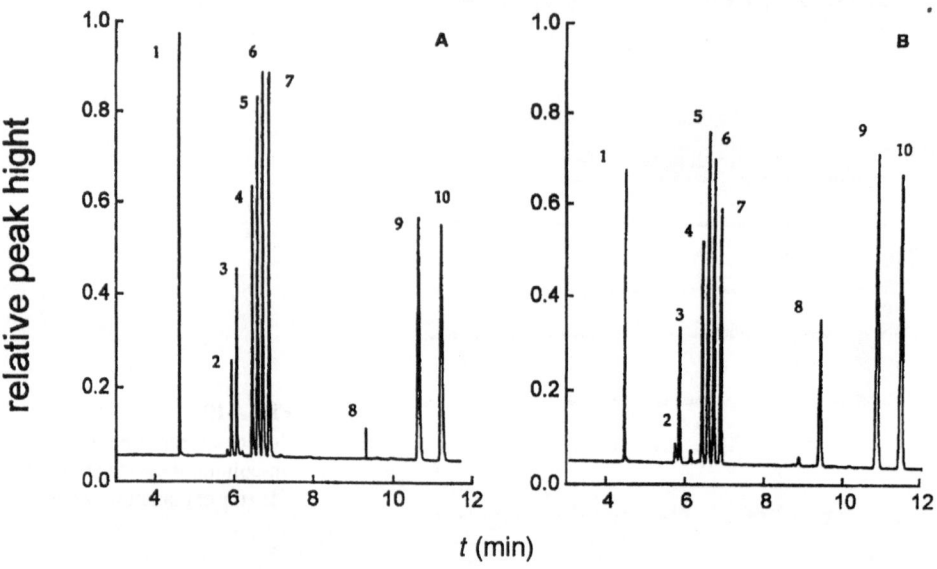

Fig. 4-20 Comparison of off-line and on-line amino acid derivatization with OPA [29a]

(A) Derivatization off-line: reaction time 10 min
(B) Derivatization on-column: reaction time 10 min after injection
Conditions: L: 66.5/73 cm; 75 µm i. d.; buffer: 40 mM borate pH 9.5
containing 0.1 mg mL^{-1} OPA and 0.1 % (v/v) b-mercaptoethanol
Detection: LIF Ex 351 nm, Em 450 nm
Analytes: Arg(1), Lys(2,3), Tyr(4), Phe(5), Thr(6), Ser(7), L-Dopa(8), Glu(9), Asp (10)
analyte concentrations: 5 µg mL^{-1} except Lys (50 µg mL^{-1}) and Asp, Glu (12 µg mL^{-1})

lary (B). In this special case the reagent was added to the buffer directly because separation and reaction conditions required the same high pH value. The performance of on-column derivatization does not necessarily require an ITP step.

An interesting variant for carrying out on-column reactions is EMMA (electrophoretically mediated microanalysis) [31]. Here, reagent and sample are introduced sequentially, the more rapidly migrating component (usually the sample) followed by the slower component (usually the reagent). If both components pass each other, a reaction occurs in the capillary. The reaction product, in turn, has a different migration velocity and can therefore be detected separately. In this way, enzymes (the capillary containing the substrate) or ions can be determined by the use of selective reagents.

The first experimental arrangements for post-column derivatization have been described, but are not as yet suitable for utilization outside research laboratories [30].

Table 4-7 Detection techniques in capillary electrophoresis

Detection principle	Detection limits absolute amount [mol]	concentration [mol L^{-1}]	Typical application	Remarks
UV absorption	10^{-15}-10^{-13}	10^{-7}- 10^{-4}	Aromatic compounds, proteins, nucleic acids,...	Currently standard detection in CE integrated into all commercial instruments
Indirect UV absorption	10^{-16}-10^{-13}	10^{-8}-10^{-4}	Metal ions, amines, organic and inorganic ions, sugars	Possible with commercial UV instruments
Fluorescence	10^{-18}-10^{-13}	10^{-9}-10^{-4}	Derivatized amino acids, DNA, peptides, proteins	Derivatization required for many samples
Laser fluorescence	10^{-21}-10^{-17}	10^{-13}-10^{-7}	DNA fragments, derivatized amino acids	Lasers are still very expensive, useable almost only in VIS or near UV range
Indirect fluorescence	10^{-16}-10^{-14}	10^{-7}-10^{-5}	Alcohols, amines, anions, cations, sugar	Only few known applications
Amperometry	10^{-16}-10^{-14}	10^{-8}-10^{-6}	Easily reducible and oxidizable substances, e.g., neurotransmitters	Suitable for capillaries down to 2 μm i.d.
Conductance	10^{-18}-10^{-16}	10^{-7}-10^{-5}	Ionic samples, e.g. metal ions, amines, carboxylic acids capillaries	Disadvantage: difficult to change
Potentiometry	10^{-19}	10^{-8}	Alkali and alkaline earth ions, selective detection by ion-selective microelectrodes	Disadvantage: microelectrodes hard to handle and prepare
Mass spectrometry	10^{-17}	10^{-8}	Proteins, peptides, drug monitoring commercial interfaces available	Unproblematic coupling,
On-line radioactivity measurements	10^{-18}-10^{-16}	10^{-10}-10^{-8}	P^{32} and C^{14} in biochemically relevant compounds	Good detection sensitivity through stop-flow systems
Flame photometric detector (FPD)	10^{-13}-10^{-11}	10^{-5}-10^{-3}	ionic organic compounds	less sensitive than comparable GC detectors

Table 4-7 contains the attainable detection limits of the most used detection systems. It shows that excellent mass sensitivities indeed can be achieved on account of the small volumes used. On the other hand, the concentration sensitivities lie in the usual HPLC range.

4.6 Special Problems of Quantitative Analysis in CE

As has already been mentioned several times, Gaussian peak shapes are seldom obtained in CE; instead, they are more or less triangular signals. As already discussed, differences in the conductances between the separation electrolyte and sample solution, large concentration differences between the individual sample components, differences between the mobilities of the buffer and sample components, and salt contents of the sample matrix lead to strongly asymmetric (tailing or leading) peaks. Since the usual measuring devices are geared to chromatographic peaks that are essentially Gaussian in shape, problems arise in the integration and the determination of the maxima of the peaks. An exact relationship between peak height and area exists only for a Gaussian curve, and does not hold for the often strongly asymmetric CE peaks. Whereas the peak maximum and center of gravity coincide for Gaussian curves, and therefore no problems arise in using the migration time for the qualitative identification, for asymmetric peaks the first moment (passage of the center of gravity) would have to be used for specifying the migration time.

Because of the peak symmetry and the more or less triangular peak shape, different results are obtained depending on whether the evaluation is performed on the basis of the peak height or area. In the quantitative evaluation of the area, it should also be noted that the zones move past the detection window at different velocities. The area (integration of the signal height with time) in CE should therefore be standardized with respect to the migration time. These special problems will be discussed separately.

Meanwhile, the miniaturization of sample introduction in commercial instruments has progressed so far that starting with a total sample volume of 3 μL, several injection can be made of the same sample in automated instruments. Changes in the sample concentration in the reservoir between injections cannot be ruled out, however. The causes of this are the contamination of the sample with the buffer and the selective injection of sample components during electrokinetic sample introduction. Working with such small volumes is also problematic because the sample volume may be altered by evaporation. This phenomenon can be reduced by cooling the sample and through completely tight seals on the sample containers. On the other hand, the advantages of CE lie in areas where only small sample volumes are available, e.g., ion analysis of a rain drop or of biological specimens.

If the raw analytical data are present in the form of a chromatogram or pherogram, quantitative results may be obtained by determining the peak height or integrating the peak area. Only in the trace concentration range is the quantitative expression in terms of peak height superior to peak area. There are two reasons for this: first, only in the trace range is the peak height proportional to the concentration inasmuch as saturation and overloading effects can be eliminated, and second, the errors resulting from automated height determination are smaller in these cases than those from integrating the peak area. Integration becomes practically necessary only when the peak heights rise.

This is demonstrated by the calibration curves of sodium for the determination of metal ions by indirect UV detection. From Fig. 4-21 it is evident that the linear range for evaluation via peak height is very short and the curve already reaches the saturation range at about 4 ppm. In contrast, a plot of peak area *vs.* concentration yields a linear correlation over a broad range. One reason for the nonlinearity with peak heights is the triangular peak shape. As has already been discussed and as can also be derived mathematically, only for Gaussian-shaped peaks is the concentration proportional to both the height and area. For other peak shapes integration is essential in order to obtain linear calibration curves [32].

As a result of the different migration velocities of the analytes passing through the detector, variations in peak areas are obtained for solutes with the same molar absorptivity. This can be illustrated for UV detection if the time is calculated during which individual substances with different velocities are visible in the detection region and their peak widths are correlated at constant recorder speeds. Such values are summarized in Table 4-8. The sample components that migrate first through the detector

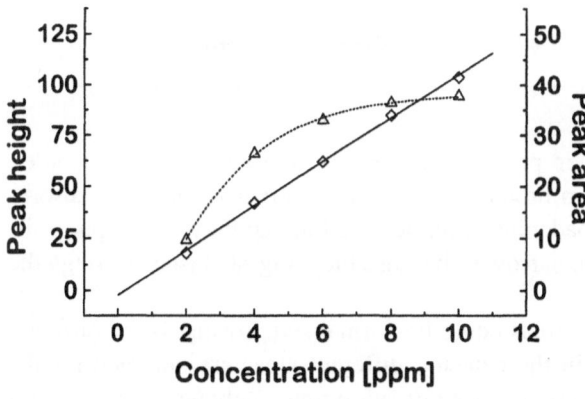

Fig. 4-21
Calibration curves for peak heights (triangles) and peak areas (diamonds) [32]

Separation conditions: CE instrument: Millipore Waters Quanta 4000; capillary: 75 μm, 50/56 cm; field: 446 V/cm; buffer: 5 mmol L^{-1} imidazole/sulfuric acid pH 5.3; injection: hydrostatic, 30 s; detection: indirect, 214 nm; sample: potassium, sodium, barium, calcium, magnesium, and lithium with concentrations of 4, 6, 8, and 10 ppm.

Table 4-8 Time of peak detection as a function of the migration velocity

Velocity of the peaks [mm/s]	Time of peaks in "window" [s]	Width of peak on recorder [mm]
0.1	20	32
0.5	4.0	6.4
1.0	2.0	3.2
5.0	0.4	0.6
10	0.2	0.4

Chart speed: 100cm min^{-1}, peak width: 1 mm, capillary length 1 m

Table 4-9. Dependence of the peak area and peak width on the migration time

Migration time [s]	Migration velocity [mm s^{-1}]	Time in the detector [s]	Peak width on the recorder [cm]	Peak area of a signal with h=10 cm [cm^2]
90	0.21	0.87	0.144	1.44
120	0.28	1.16	0.192	1.92
150	0.35	1.44	0.239	2.39
300	0.70	2.88	0.478	4.78

possess a high velocity and their peak width appears narrower on the recorder, whereas those detected later seem broader even though their mass and UV absorption are the same. If no band broadening at all occurred in capillary electrophoresis, the initial peaks would still be the narrowest because they migrate fastest through the detector.

Table 4-9 indicates how peak areas could be normalized. Owing to the different residence times of the analytes in the detector, different areas are obtained for the same amounts of sample as a function of the migration time. Only when the peak areas are divided by the sample migration times is a constant value obtained. The same result is obtained if the peak area is standardized with respect to the residence time in the detector. However, the latter is not readily achieved experimentally. Through this normalization, more reproducible quantitative results are obtained even for fluctuating migration times.

The reproducibility of the migration times is influenced less by the electrophoretic conditions than by the fluctuations of the EOF during the analysis. Irreversible adsorption of sample components, buffer changes, etc. affect the EOF very strongly and concomitantly, the reproducibility of the migration times. These exert similar effects on the reproducibility of the analytical results. Fig. 4-22 shows the variation of the migration times of some test substances over the course of the first 27 injections in a new capillary. Following an equilibration phase of about 4 injections, a substantially constant value was attained, with only minor fluctuations in the migration times. After the ninth injection the capillary was stored for 48 hrs in the separation buffer. After a short preliminary phase, a larger EOF is attained. This plateau remained constant for the subsequent 20 injections.

These experiments illustrate that reproducible analysis conditions can be attained after a short equilibration time. After a change in buffer, the capillary should be changed or the old one flushed at least 10 to 15 min with the new buffer. Since the EOF is substantially more effective in exchanging the buffer layers along the inner capillary surface than pressure-driven flow, the equilibration times may be shortened if the capillary is flushed with the new buffer and then driven by an applied potential for 5 to 10 min. Exact times for the equilibration of the capillary cannot be given because the values are strongly dependent on the type of buffer and the individual conditioning steps. It seems more practical and less time consuming to always use the cap-

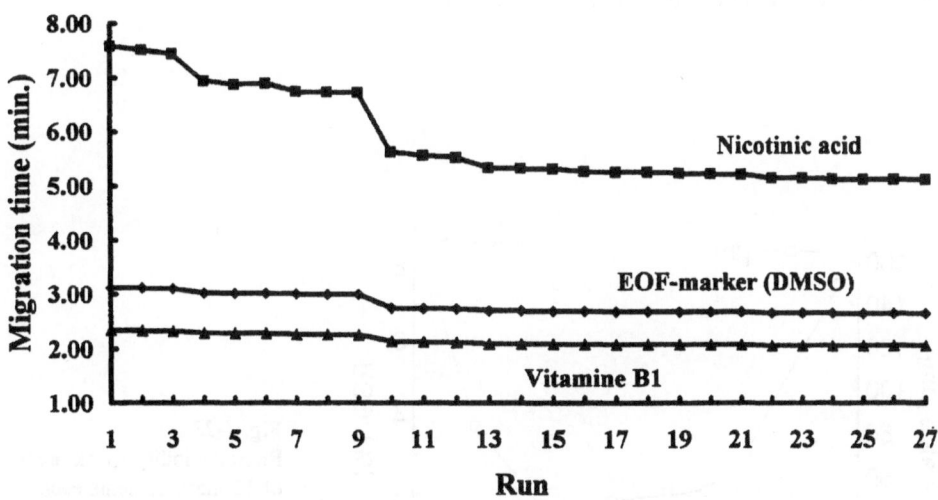

Fig. 4-22 Reproducibility of the migration times of an unused and uncoated capillary

Separation conditions: L=40/47 cm, 50 μm i.d., buffer: 10 mM phosphate pH 7.0, potential: 20 kV, detection at 214 nm, flush time between each run: 2 min with 0.1 M NaOH, 3 min with buffer

illary for only one buffer or application. The capillary should always be stored in the buffer and measures have to be taken to prevent the capillary from drying out.

Furthermore, as in chromatography, the reproducibility is affected by the amount of sample injected. If one works in the range of the detection limit, the errors in quantification are considerably higher even though that is where the highest separation efficiency is attained. Fig. 4-23 shows the relative standard deviation for quantitative analysis and the plate height obtained as a function of the amount of sample injected. In this series of measurements 12 analyses were carried out at each of the various sample concentrations and the peak areas determined. Prior to each new sample concentration, fresh buffer was placed in the containers. Statistical evaluation revealed clearly that at concentration 20- to 50-fold above the detection limit, peak integration afforded good results. The reproducibility was in the range of 2 - 3%. For integration near the detection limit the errors rose rapidly to over 7%. In this case the efficiency, and hence the ability to separate adjacent peaks, were better. Efficiency drops with increasing concentration and reaches an H-value of only 140 μm for a sample concentration of 100 mmol L^{-1} when the system is overloaded. The large increase in the H-value can be attributed to the appearance of triangular peaks at high concentrations.

In optimizing an analysis it is necessary to compromise between reproducibility and efficiency. To obtain reliable quantitative results, the sample concentration should be 10 to 20 times above that of the detection limit. High separation efficiencies and good reproducibility of the peak areas in the range of 1.5 to 3% can be achieved only under optimized conditions.

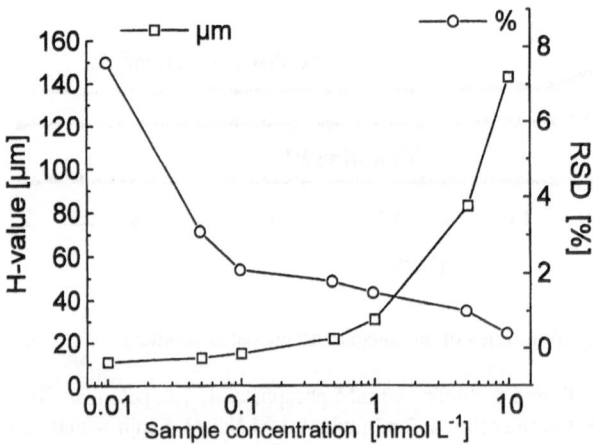

Fig. 4-23
Reproducibility (peak areas of 12 measurements each) and band broadening as a function of the sample concentration

Thus, the reproducibility of quantitative analyses depends on numerous individual factors. Comparative measurements with various CE instruments in different laboratories demonstrated that the methods are basically transferable; however, the analytical accuracy fluctuates between 1 and 2.5%. The results of a recently published inter-laboratoty comparison (Table 4-10) shows that more emphasis should be placed on individual errors than those attributable to instrument and method [33].

Table 4-11 summarizes the factors that may influence the qualitative as well as quantitative results. The various causes are given, and ways are suggested for eliminating the error sources. The principal problem in CE is the low concentration sensitivity of optical detectors that forces one to operate in the range of the detection limit. An additional problem is the restricted dynamic range of the method, as, on the one hand, the electrolyte concentration should be kept low because of Joule heating but on the other, an unfavorable concentration ratio between the ions in the electrolyte and in the sample leads to asymmetric peaks that reduce the efficiency and adversely affect quantification. In CE one is therefore often forced to dilute the samples in order to operate in the optimal concentration range. More difficult are the steps required to raise the detection sensitivity when the analyte concentrations are too low. In such cases electrophoretic methods, such as the already mentioned electrostacking or the use of isotachophoretic separation techniques prior to the actual analysis can be helpful.

Table 4-10. Comparison of reproducibility in various laboratories [33]

RSD values (n=10)							
Laboratory No.	1	2	3	4	5	6	7
Migration time	1.3	0.3	0.8	0.4	0.6	0.5	0.2
Peak height	1.0	2.3	2.1	-	1.7	1.4	1.7
Peak areas	1.2	2.6	0.6	1.3	2.2	2.5	1.7
Normalized peak areas	0.8	2.5	1.0	1.2	2.1	1.1	1.7
Linearity of the calibration curve	0.999	0.997	0.999	0.992	0.996	0.990	0.994

Measurements: Beckman P/ACE in labs. 1,2,3, and 6; ABI in labs 5 and 7;
Spectra-Physics in lab 4
RSD = relative standard deviation

Table 4-11 Overview of the factors that affect the reproducibility of peak areas

Influence by	Cause, effect	Improved by
Temperature fluctuations	- Viscosity differences - different amounts of sample injected	Thermostating the capillary and the buffer vials
Evaporation of sample solution	- Higher temperature in autosampler - raises sample concentration	Tight seals on the sample vials or cooling the sample vials
System inaccuracies	- Detector: rise time or too low data acquisition rate - Injection pressure or time too inaccurate	Use longer injection times
Peaks become larger/smaller	- Poor injection system - change in the sample after repeated injection	- Use a capillary with a flat surface cut - remove polyimide coating (2 mm) at the inlet
Poor peak shape	Wall adsorption	Coating or highly concentrated buffers
Low signal/noise ratio	Integration errors	- Optimization of integration parameters - raise sample concentration - sample stacking
Change in the migration times	- Buffer pH not constant - sample matrix adsorbed on wall - buffer level changes	- more frequent change of buffer - conditioning of the capillary with NaOH
Electrokinetic injection	Peak area becomes dependent on sample matrix	Hydrodynamic injection

5 CAPILLARY ZONE ELECTROPHORESIS (CZE)

Of the methods to be described here, zone electrophoresis is the simplest and the most extensively used. Since many of the analytical techniques to be considered are based on it, the most important principles will be discussed in detail. In zone electrophoresis the buffer, pH value, and field strength remain constant for the entire separation. The analytes are separated on the basis of their different mobilities. They are introduced in a mixture as a concrete zone at the beginning of the capillary and are sensed by the detector as discrete zones, separated from each other. In this separation method the buffer has the function of holding the pH constant and of assuring the passage of the current. The buffer pH selected determines the charge of the analyte molecules and thereby the migration direction. The buffer concentration and its pH affect the EOF. The selectivity is optimized by choosing a suitable pH, and the analysis time then depends strongly on the buffer concentration, the capillary length, and the applied potential. Further optimization of the selectivity can be achieved with buffer additives such as urea, cyclodextrin, detergents, organic solvents, etc. A schematic of the separation is presented in Fig. 5-1. Since the detector is placed on the cathode side in the conventional arrangement, cations migrate most rapidly because they move with the EOF (comigration). Anions migrate counter to the EOF (contramigration) and can be detected only if their vectorial electrophoretic migration is smaller than the velocity of the EOF. The faster their migration velocity to the an-

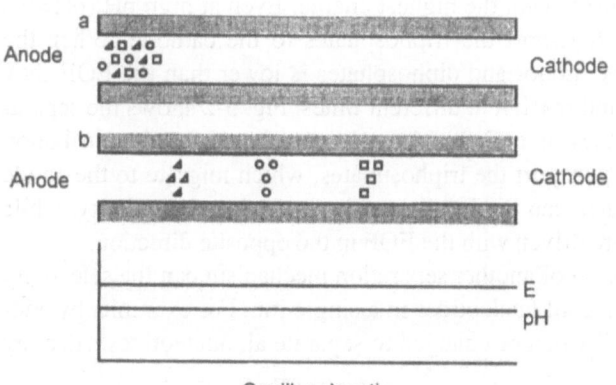

Fig. 5-1
Schematic of capillary zone electrophoresis (CZE)

ode, the later they are detected. In the extreme case when their migration velocity exceeds that of the EOF, they move out of the capillary into anode compartment and are not detected.

The resolution R in CZE can be determined from the following equation

$$R = \frac{\sqrt{N}}{4} \cdot \frac{\Delta u}{\overline{u}} \tag{5.1}$$

This equation is analogous to that of chromatography [34] if the relative migration velocities are inserted. These are proportional to the relative observed mobilities that include the contribution of the EOF because the observed mobility is comprised of the electrophoretic mobility and the EOF contribution:

$$\frac{\Delta u}{u} = \frac{\mu_{1,obs} - \mu_{2,obs}}{\mu_{obs}} = \frac{\mu_1 - \mu_2}{\mu + \mu_{EOF}} \tag{5.2}$$

It is evident from this that increasing the mobility simultaneously raises the EOF, and the relative velocity for each pair of zones is reduced along with the resolution. On the other hand, if the analyte migrates against the EOF, the resolution increases because the residence time in the capillary is prolonged and the effective migration distance is lengthened, so that even components with small differences in mobility are separated from each other. The highest resolution would be attained if the electrophoretic mobility had exactly the opposite value of the EOF. That, however, would make the analysis time infinitely long. Undissociated sample components migrate at the velocity of the EOF through the capillary. As has been shown, the EOF depends on the pH and other electrolyte properties, particularly the ionic strength. Both dependences have already been discussed. The possibilities resulting from a superimposition of the electrophoretic mobility onto the EOF can be illustrated with an example of the separation of nucleotides. The EOF is directed toward the cathode whereas the nucleotides migrate toward the anode because of their negative charge, the fastest being the triphosphates with the highest charge. Even at high pH (pH>10) the EOF is not sufficient to transport the triphosphates to the cathode. When the electrophoretic mobility of the mono- and diphosphates is lower than the EOF, they are conveyed to the detector and reach it at different times. Fig. 5-2 shows the separation of these analytes. Under these experimental conditions, the vectorial contribution of the EOF is insufficient to transport the triphosphates, which migrate to the anode compartment. The triphosphates can be determined by reversing the polarity, while the di- and monophosphates are driven with the EOF in the opposite direction.

Only through the introduction of another separation mechanism can the selectivity be altered sufficiently to resolve all nucleotides in a single run. For example, by adding ion-pair formers the mobility can be changed to separate all nucleotides. Also, by

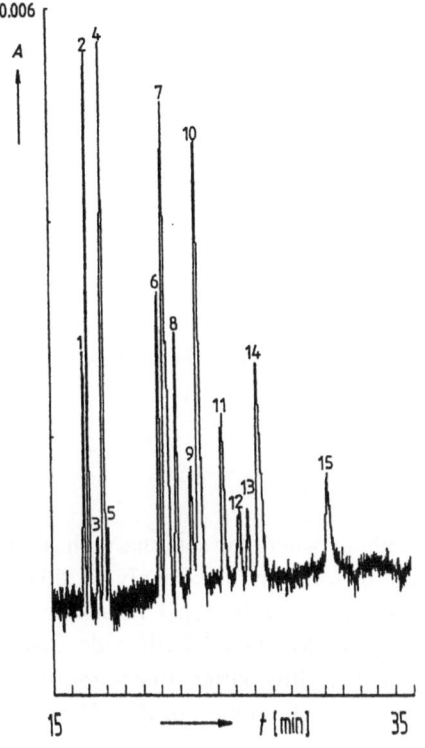

Fig. 5-2
Separation of some nucleotides by CZE [6]

Separation conditions: capillary 75 µm, 75 cm; buffer: 100 mM Chaps (3-(cyclohexylamino)-1-propionic acid) pH 10.5; field: 200 V cm^{-1}; analytes: AMP (1), dAMP (2), CMP (3), dCMP (4), dTMP (5), GMP (6), dGMP + UMP (7), dADP (8), ADP (9), dGDP (10), CDP (11), GDP (12), dCDP (13), dTDP (14), UDP (15); analyte concentrations: 100-250 µM; trinucleotides migrate faster to the anode at this EOF and cannot be detected.

reversing the EOF through surface modification and simultaneous field reversal a separation of anions with large differences in mobilities can be carried out in a single analysis. Moreover, better selectivity can be achieved through cationic micelles (*cf.* MEKC).

5.1 Principles of Optimization in CZE

5.1.1 Effect of pH

The effect of pH on the transport of the analytes to the detector may be attributed to two factors. As has already been pointed out, separations based on electrophoretic migration have the EOF superimposed on them to a greater or lesser extent, depending on the degree of dissociation of the surface silanol groups. Also, the mobility of the ions is determined by their extent of dissociation in the carrier electrolyte and therefore by its pH. Hence, a separation can be optimized by changing the pH and the type of buffer. For this, a pH range of 2 to 12 is available. Particularly advantageous

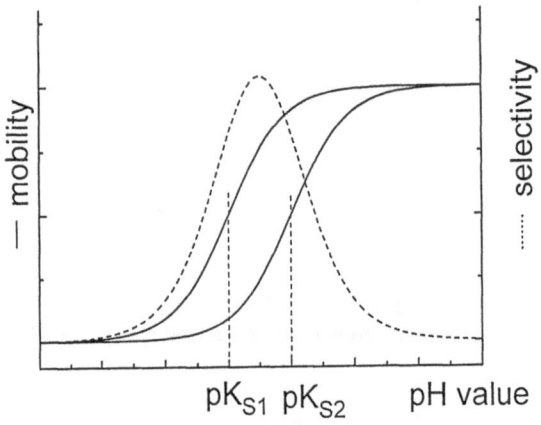

Fig. 5-3
Dependence of the mobilities of two weak acids on pH
(dashed line: selectivity)

is the alkaline range in comparison to HPLC methods where the stationary phases begin to dissolve above pH 8 and therefore cannot be used. In the strongly alkaline pH range in CE even phenols and sugars are deprotonated. Only at pH values below 2 and above 12 is the current transport overtaken by hydrogen and hydroxide ions, respectively. Due to their very high mobilities, only very low buffer concentrations can be used in these ranges. This is the sole reason for the limitation of the pH range in CE.

The largest migration differences between two basic or acidic electrolytes, i.e., the highest selectivities are obtained when the buffer pH lies between the pK values of the sample components. This behavior is analogous to that for separations by ion exchange chromatography, and is presented schematically in Fig. 5-3.

5.1.2 Effect of buffer concentration

The buffer concentration selected should be high enough to maintain the pH constant during the analysis and to keep overloading effects to a minimum, while still permitting rapid analysis via the EOF but preventing the appearance of band broadening through thermal effects. The impact of the ionic strength on the EOF has already been discussed in Section 3.3. The problems of electrodispersion were also pointed out there, as well as the difficulties associated with increasing the buffer concentration. Higher buffer concentrations can be used with smaller i.d. capillaries. For the most frequently used capillaries of 75 μm i.d., 10 to 50 mM buffers are commonly employed. Apart from modifications of the buffer system with special additives, changes in the buffer pH and concentration usually suffice to optimize the separation of charged samples. Fig.5-4 summarizes the factors to be considered when the buffer concentration is changed.

5.1.3 Buffer selection

The choice of buffer requires several factors to be considered. As has been shown, the mobilities of the analyte and buffer ions must be similar in order to achieve high separation efficiencies. The buffer concentration should be greater than that of the ions in the sample solution as this also leads to sharp, symmetrical zones because only under such conditions do the sample ions not influence the electric field. On the other hand, a high ionic strength for a given potential drop means a high current density and a rise in the Joule heating. As shown, these effects can be simply measured as deviations from Ohm's Law. Organic zwitterionic buffers (e.g., 3-(cyclohexylamino)-1-propanesulfonic acid, CAPS) have decided advantages for use with large i.d. capillaries.

In modern instrumentation the buffer vials are usually small. During electrophoresis electrolysis takes place in the buffer vials, generating protons in the anode and hydroxide ions in the cathode compartments. Therefore, sufficient buffer capacity is required to achieve reproducible analyses because the buffer is transported into the capillary by the EOF. Fig. 5-5 shows the change in pH with time for various phosphate buffers (10 mM). Calculated lines and measured values are given. Surprisingly, the largest changes are observed in the first 30 to 60 minutes after applying the potential. Only where the buffer capacities are high (at pH 3 and 7 for phosphate buffers) is the required long-term stability assured. Therefore, only high-capacity buffers should be used and the buffer solutions in the vials should be frequently replenished.

The pH determines the charge of the analyte ions and decisively changes the selectivity of the separation system. An overview of the effects of the individual parameters on the separation system and the complexity of the optimization strategies are presented in Fig. 5-6.

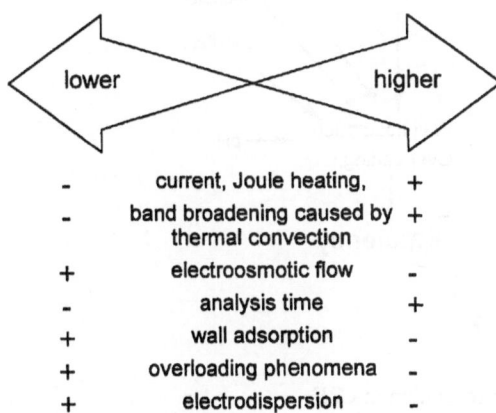

Buffer concentration

| | lower | | higher | |

−	current, Joule heating,	+
−	band broadening caused by thermal convection	+
+	electroosmotic flow	−
−	analysis time	+
+	wall adsorption	−
+	overloading phenomena	−
+	electrodispersion	−

Fig. 5-4
Effect of buffer concentration on the important quantities of capillary zone electrophoresis

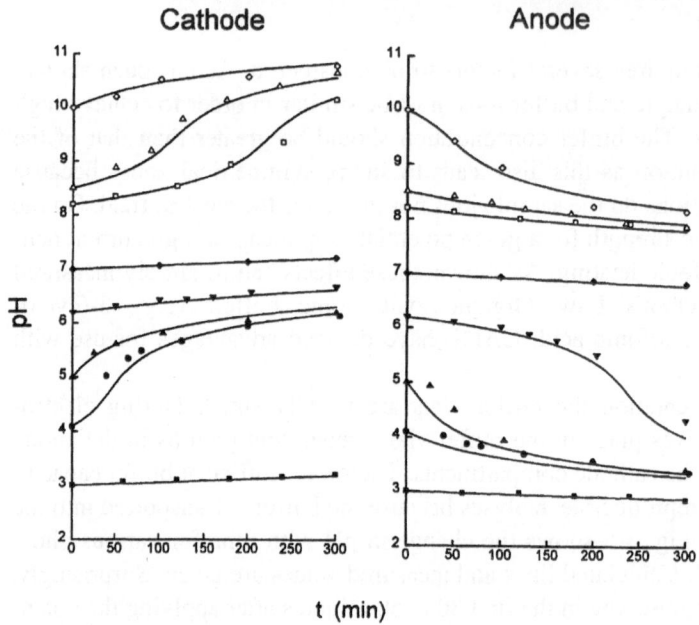

Fig. 5-5 pH changes in the buffer vials resulting from electrolysis. Conditions as in Fig. 11-1.

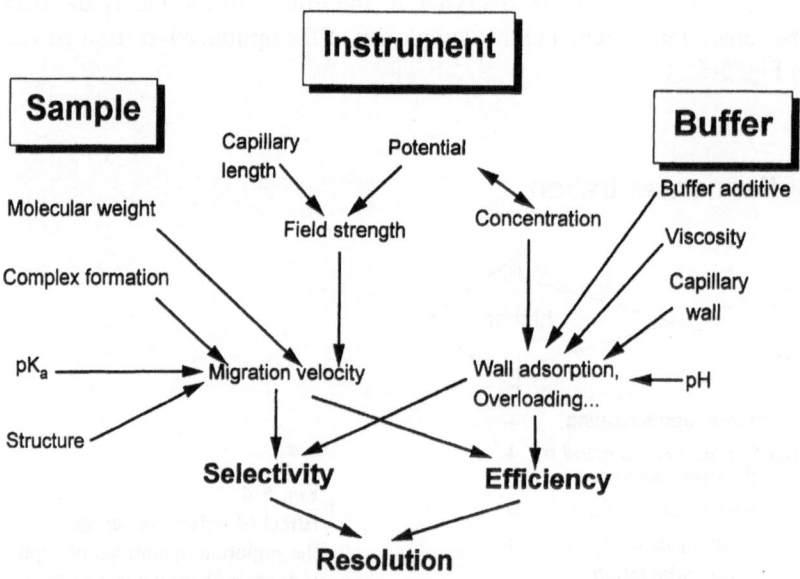

Fig. 5-6 Summary of the factors that influence separation in CZE

Since the analysis time is inversely proportional to the applied field strength, shorter analysis times can always be achieved with lower buffer concentrations. In summary, the following requirements may be placed on a buffer system in CZE:
- selectivity for the ions to be separated
- pH stabilization, buffer capacity (reproducibility)
- low UV absorbance at the detection wavelength
- similarity in the mobilities of the analyte and buffer ions
- the counterion should have a low mobility (small currents)
- reproducibility in the preparation of the buffer
- stability of the buffer

5.1.4 Applications

Electrophoresis in a continuous buffer system with a uniform field strength is the standard technique that is particularly suited for the analysis of small, permanently charged species. In this way, aliphatic and aromatic carboxylic acids, sulfonic acids, amino acids, phenols, and amines can be separated without great difficulty provided they carry suitable chromophores for detection. This method is also used for the separation of peptides and proteins, but due to the additional problems with wall adsorption, these separations are treated in a separate section (5.3). The separation of ions without a chromophore, using indirect detection techniques, will also be described in a separate section (5.2).

Fig. 5-7 shows the separation of a natural tannin mixture as an example of the efficiencies attainable in CZE. Next to the electropherogram are the UV spectra recorded on-line with a diode array detector.

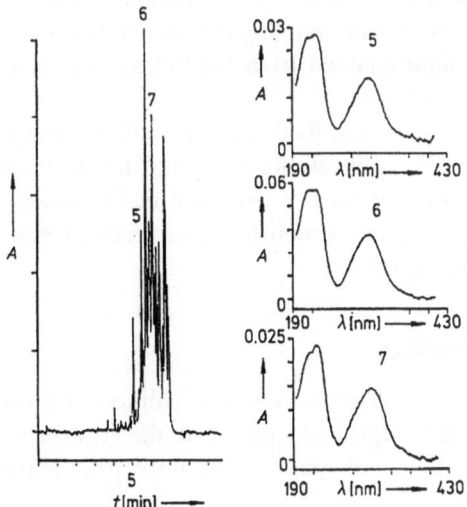

Fig. 5-7
Separation of a mixture of tannins (pyrogallic acids) [35]

Separation conditions: Capillary: 50/ 80 cm, 75 μm i.d., buffer: 50 mM borate pH 9.5; potential: 35 kV.

The separation of charged species having relatively large hydrophobic residues can be improved by the addition of SDS (sodium dodecyl sulfate). The potentials for the optimization of CE separation of nonionic compounds by the addition of detergents are discussed in detail in the section on micellar electrokinetic chromatography (MEKC).

5.2 Indirect Detection Methods in CE

For various classes of substances that possess only low molar absorptivities over the entire UV region, high concentrations must be frequently injected in order to obtain a detector signal of the compound. However, for such samples the separation system is frequently overloaded, and the peak efficiencies are so poor that practical application of such separations is not possible. Such substances can be detected considerably more sensitively by other detection methods (e.g., conductimetry and potentiometry). Since up to now, however, there are no other detectors in commercial CE instruments for routine applications, indirect UV detection assumes a special significance in CE.

This variant of UV detection has been known in chromatography since the early 80's and has been used in ion-pair chromatography and ion-exchange chromatography (IEC) for the detection of ionic compounds. In this way, non-UV absorbing substances can also be detected with conventional UV detectors. The disadvantage of this method in HPLC is the appearance of so-called system peaks that must be additionally separated from the substance zones by an optimization of the separation. A further disadvantage is the high buffer concentration required in IEC to elute the analytes from the stationary phase. The long path length in HPLC causes a high background absorption and results in increased noise. This method failed to achieve a greater breakthrough because the development of conductivity detection in HPLC provided strong competition for routine applications. The coupling of conductance detectors with and without the suppression technique enabled 10 to 100 times more sensitive detection to be achieved for routine applications.

In CE the indirect UV detection method was described early on, without recognizing its enormous significance and versatile applicability. Not until the work of Jandik and Jones [36] did the advantages of indirect UV detection in CE become clear. In addition to short analysis times, the method demonstrated surprisingly good detection limits in the range of 0.1 mg L^{-1} and lower.

5.2.1 Principles of indirect detection techniques

The Lambert-Beer Law also holds for indirect UV detection. The addition of a UV absorbing ion to the buffer generates a high background signal with the absorbance A_1. By displacing the UV absorbing buffer ions with analyte ions at the peak maxi-

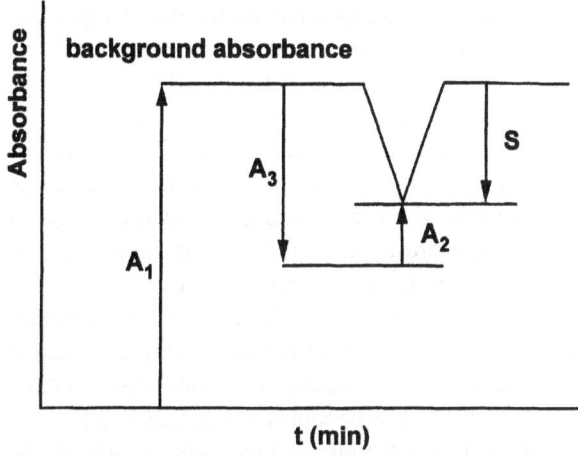

Fig. 5-8
Schematic representation of
signal generation in indirect
UV detection

mum, the background absorbance is reduced by the amount A_3, as is represented
schematically in Fig. 5-8.

To maintain electroneutrality in the electrolyte, charge displacement occurs according to the charge densities between the analyte ions and the background ab-
sorber. In the ideal case, if the buffer component is singly charged while the analyte
is doubly charged, then on purely mathematical grounds two molecules of the background electrolyte are displaced; the signal is proportional to the charge ratio between the electrolyte and analyte ions. If the analyte molecules absorb as well, the
signal is diminished by the amount of their absorbance A_2. The highest sensitivity is
attained when the analyte possesses no absorption of its own.

The magnitude of the signal for equally charged ions is the result of the difference
in the molar absorptivities between the electrolyte ε_E and the analyte ε_A, as well as
the analyte concentration c_A and the path length d

$$S = (\varepsilon_E - \varepsilon_A) \cdot c_A \cdot d. \tag{5.3}$$

Since the various detectors differ in the intensity of the light source and the type
of receiver, for the intercomparison of detectors this equation must be extended with
an individual detector constant.

For the use of indirect detection in HPLC it was found [37] that the noise according to

$$N = n \cdot c_E \cdot \varepsilon_E \cdot d \tag{5.4}$$

is proportional only to the concentration, molar absorptivity, and path length. This results in an independence of the path length for the signal/noise ratio, which for CE
would be advantageous because of its short path lengths. An experimental study
showed, however, that though this may be valid for the relatively long path lengths in
HPLC, in CE there is a constant contribution to the noise that needs to be taken into

account, and therefore that the signal/noise ratio is dependent on the path length, i.e., the capillary diameter. This constant contribution to the noise is detector-specific and is proportional to the electronic noise. Its magnitude lies between 10^{-5} and 10^{-4} absorbance units. Knowledge of this quantity permits calculation of the theoretical detection limit for indirect UV detection. It amounts to about 10 µmol L^{-1} and agrees very well with the practical detection limits. It is assumed that the actual ionic charge is known, which does not always agree with the formal charge, even for inorganic ions. Thus, it is found that Ca^{2+} ions in the presence of sulfate ions have a formal charge of 1.3 and Mg^{2+} ions under identical conditions a formal charge of 1.6.

Due to the short path lengths in CE, one can operate at the absorption maximum of the buffer component for indirect UV detection, so that measurements are made at the highest sensitivity where the signal/noise ratio is highest. The single disadvantage of indirect UV detection is the increased noise that becomes noticeable at high currents, which can be attributed to the greater thermally induced convection in the buffer solution.

Since fundamental differences exist between the analysis of cations and anions with indirect UV absorption that concern the choice of the UV absorbing buffer components and the modification of the EOF, the two separation systems will be described separately.

5.2.2 Separation of cations with indirect UV detection

The first separation of cations using indirect UV detection was described by Foret *et al.* [38] in 1990. It was shown that CE is superior to ion chromatography for the separation of the lanthanides in terms of the separation efficiency, analysis time, and peak resolution. All lanthanides could be separated by CE, which is not so readily achieved even in optimized ion chromatographic systems.

Theoretical consideration show that in fused silica capillaries the electroosmotic flow and the migration of cations is always in the same direction. The CE apparatus is used with normal polarity where the end of the capillary is grounded. In this way, the superimposition of the ion transport by the EOF through the electrophoretic migration of the cations permits very rapid separations.

As has already been pointed out, the buffer components should possess a mobility similar to that of the ions to be separated. If the sample components migrate faster than the displaced buffer ion, leading peaks are obtained; if they migrate more slowly, peak tailing occurs. This is demonstrated in Fig. 5-9 for potassium and lithium ions with imidazole as background electrolyte. Only the calcium peak is reasonably symmetric because its mobility is similar to that of imidazole. The most important optimization step for the separation of cationic compounds is the choice of a suitable cationic background electrolyte.

Table 5-1 affords an overview of possible buffer components for the mobility range of 0.17 to 0.55 $cm^2 kV^{-1} s^{-1}$. These components cover the entire mobility range

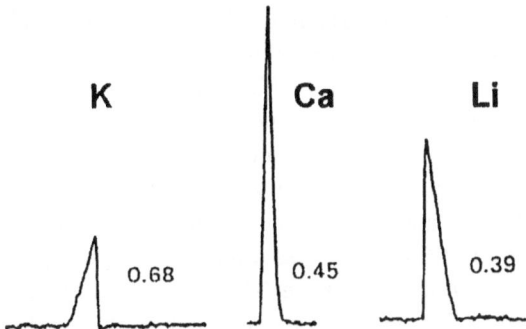

Fig. 5-9 Effect of mobility differences between analyte ions and the background electrolyte on peak symmetry (mobilities in $cm^2 V^{-1} s^{-1}$)

Table 5-1 Overview of important characteristics of cationic buffer substances

Substance	Mobility [$cm^2 kV^{-1} s^{-1}$]	pK_a	useful spectral region
Histamine	0.53	6.5; 10,5)[1]	< 220 nm
p-Aminopyridine (PAP)	0.48	8.9)[1]+)[2]	< 220 nm and 240 - 270 nm
Imidazole	0.46	6.9)[1]	< 220 nm
9-Aminoacridine	0.42	9.2)[2]	250 - 265 nm
Creatinine	0.37	5.3)[2]	240 - 270 nm
2-Amino-4,6-dimethylpyrimidine	0.32	8.1)[2]	< 220 nm and 265 - 280 nm
Ephedrine	0.31	9.9)[2]	< 220 nm
4-Amino-N,N-diethylaniline	0.27	8.1)[2]	< 215 nm
2-Aminobenzimidazole	0.20	7.5)[2]	< 235 nm and 270 -310 nm
1-Aminonaphthalene	0.17	39)[1]	< 230 nm, about 300 nm

[1]: Handbook of Chemistry and Physics, 67th Edition, CRC Press, Boca Raton 1986 (p.159)
[2]: Determined by titration

of inorganic ions, as well as amino acids, amines, etc. The table also gives the optimal wavelength region for UV detection. For the separation of the alkali and alkaline earth ions as well as aliphatic amines, imidazole has proved itself to be the best. Its high absorption at low wavelengths, however, limits its applications to ions with no absorbance between 200 and 220 nm. As was shown, the signal intensity decreases for ions that absorb in this range. As an alternative, p-aminopyridine (PAP) has proved its value and can be used at pH's above 6 where imidazole is scarcely protonated. At its absorption maximum of 205 nm PAP has a molar absorptivity of 15,500, at 245 nm about 15,600, and still 14,000 at 254 nm. This is about triple that of imidazole at 214 nm. The mobility of PAP is nearly identical to that of imidazole. But PAP offers advantages because of the large wavelength range, so that it can be used for indirect detection even at 270 nm. Moreover, PAP buffers can be used up to pH 10.1 without loss in sensitivity due to a decrease in the charge. Trimethyl-phenylammonium bromide with a mobility of 0.36 cm^2 kV^{-1} s^{-1} is used frequently but has only a very low molar absorptivity that prevents achieving the detection sensitivity of imidazole or PAP buffers. Creatinine has limited solution stability and storage under a protective gas is recommended.

Fig. 5-10 illustrates the advantages of an imidazole buffer system, having a concentration of about 5 mM and a pH below 6. The separation is completed in less than

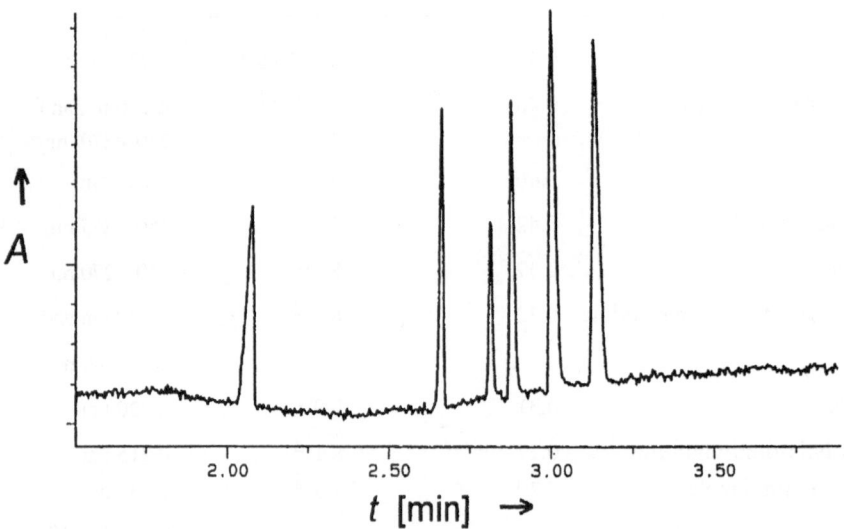

Fig. 5-10 Separation of alkali and alkaline earth ions in an imidazole buffer system [32]

Separation conditions: CE instrument: Millipore/Waters Quanta 4000; capillary: 75 μm, 50/56 cm; field: 446 V cm^{-1}; buffer: 5 mmol L^{-1} imidazole/sulfuric acid pH 4.0; injection: hydrostatic 30 s; detection: indirect 214 nm; sample: cation standard with 1 mg L^{-1} of potassium, sodium, magnesium, barium, calcium, and 0.5 mg L^{-1} of lithium

4 min. After about 5.5 min a large peak appears in the electropherogram (not shown here), which is caused by the electroosmotic flow and may be attributed to the injected water. The water has a lower UV absorption than the buffer system and therefore gives rise to a negative peak from which the electroosmotic flow can be determined.

The buffer concentration has a great effect on the peak shape: The peaks become more symmetric with increasing concentration. The plate number quadruples when the buffer concentration is raised from 0.5 to 12 mM. There are clear limits to this optimization step, however, as the noise begins to increase from a buffer concentration of 8 mM. This sharp rise is accompanied by Schlieren effects in the buffer, which are caused by a density gradient that, in turn, results from temperature changes due the Joule heating in an electric field. With 100 µm i.d. capillaries imidazole buffer concentrations up to 5 mM can be used; for 75 µm ones up to 10 mM in the pH range 4-6, and for 50 µm ones up to 50 mM.

Optimization with respect to analysis times can be readily achieved. By means of low concentration buffers (2 mM) and a pH of 6.0 and high field strengths a large

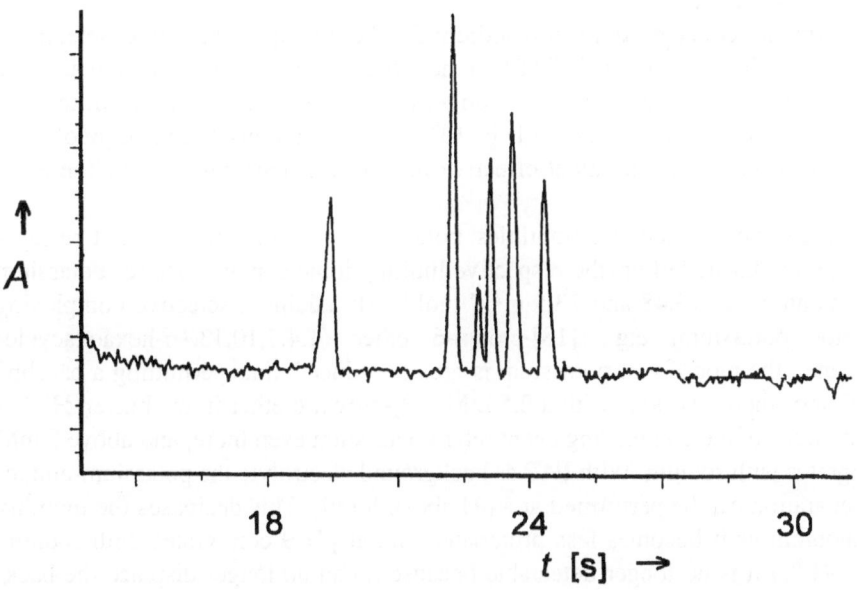

Fig. 5-11 Time-optimized separation of test ions in an imidazole buffer system [32]

Separation conditions: CE instrument: Millipore Waters Quanta 4000; capillary: 75 µm, 20/26 cm; field: 1304 V cm^{-1}; buffer: 2 mmol L^{-1} imidazole/sulfuric acid pH 6.0; injection: hydrostatic, 10 s; detection: indirect 214 nm; sample: cation standard with 1 mg L^{-1} of potassium, sodium, magnesium, barium, calcium, and 0.5 mg L^{-1} of lithium

electroosmotic flow can be generated. A short separation distance leads to extremely fast analyses. As Fig. 5-11 illustrates, the test ions were separated in less than 30 sec with a field of 1300 V cm^{-1} (30 kV), with outstanding peak efficiency despite the very high field. For example, for calcium 600,000 plates per meter were attained, or normalized with respect to time, 320,000 plates per min. The high field strength accelerated the ions to a migration velocity of 0.96 cm s^{-1}, without appreciably changing the flow profile and thus the peak efficiency.

The selectivity can be varied by the addition of other ions and complexing agents. Although it cannot be predicted from theory, the separation of cations also depends on the counter ion that is introduced with the background electrolyte. Sodium and calcium can be separated without difficulty with sulfate and phosphate in an imidazole buffer, but resolution decreases when these anions are replaced with perchlorate, chloride, or acetate. With trifluoroacetate as counter ion no separation is possible of Na-Ca in an imidazole buffer. The interaction of ions with complexing agents can be used to selectively modify their mobilities. Although the addition of oxalate ions to the buffer had no effect on the separation of alkali and alkaline earth ions, citrate strongly reduced the mobilities of the alkaline earth ions. For example, 100 µmol L^{-1} citrate was sufficient to retard calcium and magnesium so that they were detected after lithium.

Even smaller concentrations of disodium EDTA are required to effect selectivity changes. Less than 20 µmol L^{-1} EDTA in the buffer suffices to complex the alkaline earth ions so strongly that they can no longer be detected. The mobilities of the singly-charged ions are affected very little by EDTA concentrations up to 200 µmol L^{-1}. EDTA has proved to be the most effective reagent for masking the alkaline earth ions.

Because of nearly identical mobilities, potassium and ammonium cannot be separated in an imidazole buffer (the respective limiting ionic conductance for potassium and ammonium are 73.48 and 73.5 m^2 S mol^{-1}). By adding a selective complexing agent for potassium, e.g., [18]-crown-6 ether (1,4,7,10,13,16-hexaoxacyclooctadecane), the mobility of potassium is greatly reduced, thus permitting a baseline separation of these two ions with a 2.5 mM [18]-crown-6 ether in the buffer. Higher concentrations of the complexing agent retard potassium even more, and above 8 mM it comigrates with sodium. With PAP as background electrolyte the potassium-ammonium separation can be performed at a pH above 8 [39]. This decreases the mobility of ammonium as it becomes less protonated and at pH 9 comigrates with sodium. Above pH 9.5 it is no longer detectable because it can no longer displace the background electrolyte.

The addition of organic solvents to the buffer has no effect on the detection of the components by indirect UV detection, even though both the EOF and the mobility of the ions are changed owing to variations in their hydration shell. Thus, by adding 20% methanol the mobilities were reduced by about 30%, and somewhat more with THF and isopropanol, whereas acetonitrile had a smaller effect. However, differences

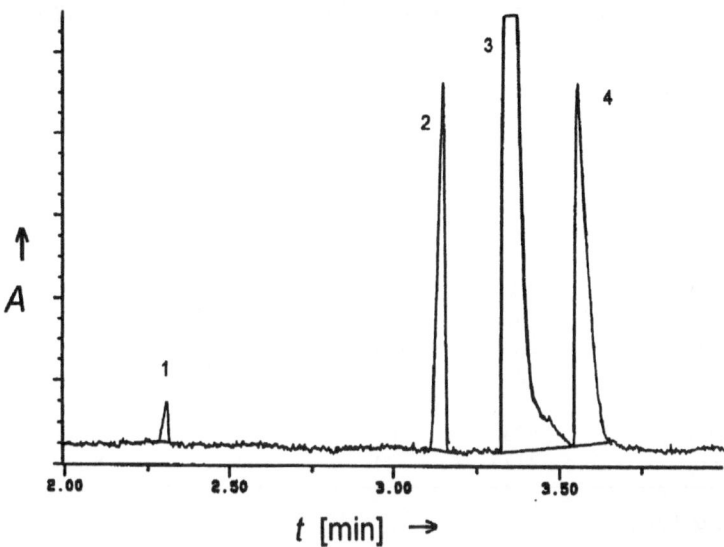

Fig. 5-12 Cation separation in surface water (Lake Baikal) [32]

Separation conditions: CE instrument: Millipore Waters Quanta 4000; capillary: 75 µm, 54/60 cm; field: 446 V cm^{-1}; buffer: 5 mmol L^{-1} imidazole/sulfuric acid, pH 4.5; injection: hydrostatic 30 s; detection: indirect, 214 nm; sample: water of Lake Baikal with : 1 potassium, 2 sodium, 3 calcium, 4 magnesium; quantitative results in Table 5-2

in selectivity were not observed. The effect of organic components on the EOF is not uniform and has already been discussed in Section 3.3.

The detection limit of the method depends on the molar absorptivity of the background electrolyte. A detection limit of 0.2 ppm was determined for PAP. With imidazole the detection limit was an order of magnitude worse. Good agreement was obtained between the calculated and measured values. The CE of cations presents a good alternative to non-suppressed ion chromatography. The analysis times for CE are shorter and the costs of capillaries and buffer are inconsequential. Moreover, the same capillary can be used for anion analysis after a buffer change. Fig. 5-12 shows the cation analysis of the water of Lake Baikal and the quantitative results are summarized in Table 5-2. The values determined by CE are compared to those obtained by other methods. Results are also presented for a low-salt water sample (Waldquelle). These examples of applications demonstrate that the results of different analytical methods agree well within certain tolerance limits.

A substantial advantage of the CE method is that it frequently requires no sample preparation. It is entirely possible to inject suspensions without filtration, which makes this method ideally suited for the determination of ions in waste water, etc.

Tabelle 5-2 Quantitative results by CE compared to IEC and AAS (all concentrations in ppm)

Sample	Method of analysis	Potassium	Sodium	Calcium	Magnesium
Water of	HPLC	0.5	5.3	15.9	3.3
Lake Baikal	CZE	0.6	3.1	14.5	2.9
	AAS	1.1	3.4	14.6	3.3
Spring	HPLC	1.2	3.2	4.7	0.8
water	CZE	1.6	3.2	4.1	1.1
	AAS	2.6	3.4	4.5	0.6
	1)	2.0	3.6	3.8	0.7
Apple vinegar	HPLC	925	69	92	52
	CZE	1125	74	114	64

1) Data from the bottle labels.

The analysis of cations in blood is more problematic. The size of the red corpuscles (7.5 x 2 μm) lies at the limit of what can be injected without causing problems in CE. After a 100-fold dilution of fresh blood, the sample could no longer coagulate. The red and white corpuscles in blood do not permit a direct injection in HPLC. The protein content of this diluted solution is still over 0.7 g L^{-1}, which makes it a substantial matrix component. Nevertheless, the cations could be separated by CE without difficulty because of their very high mobility in comparison to that of the matrix components, as is shown in Fig. 5-13 [40].

Due to the good detection limits, these ions could be detected even in this dilute solution. The concentration values obtained by comparison with externally constructed calibration curves agree well with literature values. For potassium 6.8 mval L^{-1} was found (literature value 5 mval L^{-1}), for sodium 147 mval L^{-1} (lit. 150 mval L^{-1}), for calcium 3.4 mval L^{-1} (lit. 2 mval L^{-1}), for magnesium 5.4 mval L^{-1} (lit. 3 mval L^{-1}).

The proteins in the solution do, however, interfere with the reproducibility of the separation. The silanol groups at the capillary wall that are responsible for the electroosmotic flow are increasingly blocked by the adsorption of the proteins from each injection. This slows the EOF and also retards the migration of the ions to the detector. The decreased migration velocity directly affects the peak areas. As the migration time increases the ions pass the detector more slowly, which increases the peak area. Because of the steady increase in the migration times, their standard deviation is in excess of 30%, that of the peak areas is 6%.

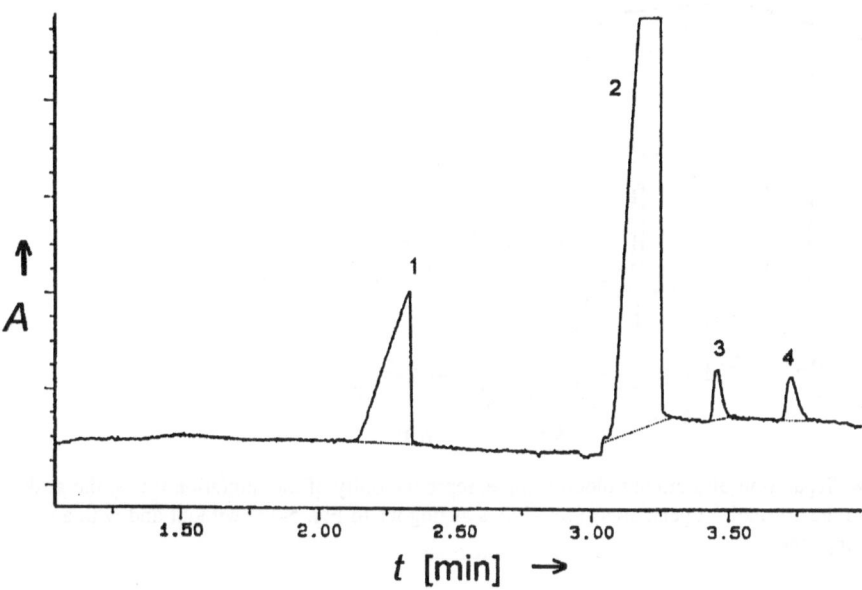

Fig. 5-13 Separation of alkali and alkaline earth ions from a diluted blood sample [40]

Separation conditions: CE instrument: Millipore/Waters Quanta 4000; capillary: 75 μm, 50/57 cm; field: 446 V cm^{-1}; buffer: 5 mmol L^{-1} imidazole/sulfuric acid; injection: hydrostatic, 30 s; detection: indirect 214 nm; sample: 50 μL blood suspended in 5 mL water; 1 potassium, 2 sodium, 3 calcium, 4 magnesium

Normalization of the peak area with respect to the migration time is simply a calculational manipulation of the analytical results. The reason for the trend in the migration time remains intact. This normalization improves the standard deviation of peak areas from 6.2 to 2.8%, as can be seen from the block diagram in Fig. 5-14. A more suitable solution to this problem, however, is the conditioning of the capillary prior to each run. Here, only a single conditioning step, rinsing the capillary with buffer for 3 min, had been used in the analyses presented. To remove the proteins in the capillary after each separation, in a second series of measurements it was flushed for 5 min with 0.1 M NaOH and for another 5 min with buffer. These steps enabled the interfering substances from the sample to be removed from the wall and then to recondition the capillary reproducibly with new buffer. The characteristic curve of migration time/peak areas exhibited no drift under these conditions, the times and peak areas fluctuating purely statistically. The reproducibility of the calcium peak area after 8 injections lay within 2%. If the area was normalized with respect to time, the value was 1.5%. Such conditioning steps are absolutely essential for inhomogeneous samples.

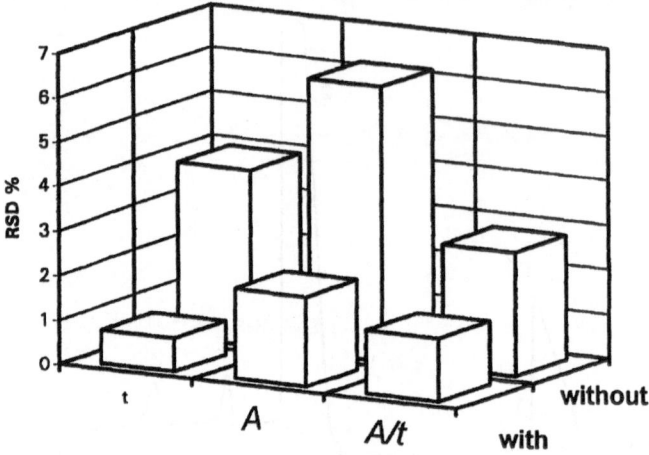

Fig. 5-14 Separation of a diluted blood sample, reproducibility of the migration times, the peak areas, and the normalized peak areas (conditions analogous to Fig. 5-13) without and with a re-conditioning step

Organic amines, morpholine, and other organic bases can be separated after protonation analogous to metal and ammonium ions. Typical pK_a values for the aliphatic amines range from 9.5 to 10.8. An illustrative separation of metal ions, amines and amino alcohols is presented Fig. 5-15. Indirect UV detection was used in conjunction with an imidazole buffer system. As a result of the high peak efficiency, all substances were baseline separated.

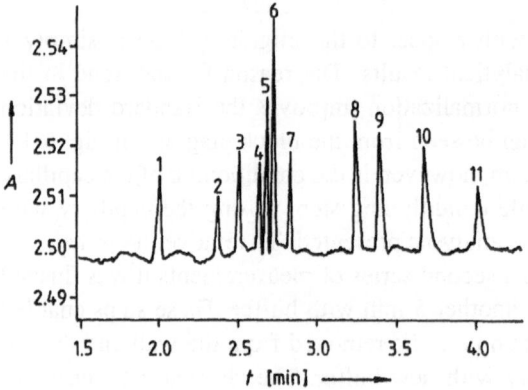

Fig. 5-15
Separation of small amines and metal ions [32]

Separation conditions: CE instrument: Millipore Waters Quanta 4000; capillary: 75 μm, 50/58 cm; field: 436 V cm^{-1}; buffer: 5 mmol L^{-1} imidazole/sulfuric acid pH 4.7; injection: hydrostatic 30 s; detection: indirect 214 nm; sample: 1.0-2.5 ppm potassium 1, sodium 2, dimethylamine 3, trimethylamine 4, calcium 5, magnesium 6, lithium 7, diethylamine 8, triethylamine 9, diethanolamine 10, triethanolamine 11

Table 5-3 Detection limits and upper limits of the linear range in direct and indirect detection modes.

Detection	direct UV 200 nm		indirect UV 254 nm	
Substance	DL	linear range	DL	linear range
Ethylamine	50 ppm	2000 ppm	1 ppm	50 ppm
Diethylamine	50 ppm	2000 ppm	1 ppm	70 ppm
Triethylamine	20 ppm	500 ppm	1 ppm	100 ppm

Separation conditions: CE instrument: Beckman P/ACE 2000; capillary: 75 μm, 50/57 cm; field: 439 V cm^{-1}; injection: pressure, 8 s; buffer: direct UV detection: 50 mmol L^{-1} borate pH 9.5, indirect UV detection 5 mmol L^{-1} ephedrine/sulfuric acid pH 7.5; detection: direct UV detection 200 nm, indirect UV detection 214 nm; sample: amines in water

Table 5-3 shows a comparison of the detection limits for direct and indirect UV detection, as well as the upper limits of the linear range for some weakly UV-absorbing amines. Similar measurements with alkali or alkaline earth ions could not be carried out because their UV absorbance is too low. It can be seen from the data that organic amines can be detected by the direct mode at 200 nm. The analysis can be performed within 5 min in a borate buffer. By using indirect UV detection at 254 nm with an ephedrine buffer, the detection limit is reduced *ca.* 50-fold to about 1 mg L^{-1}. The linear range for direct UV detection extends about 1.4 to 1.6 orders of magnitude and 1.7 to 2 orders of magnitude for indirect UV. This demonstrates that the linear range of indirect UV detection can also be fully utilized, and that CE with indirect UV detection for precisely such samples complements chromatographic separation systems.

5.2.3 Separation of anions with indirect UV detection.

The separation of anions by CE in the normal configuration (with the detector on the cathode side) is more difficult because the anions migrate against the EOF. Although higher selectivities are obtained because of their contramigration, as Fig. 5-16 shows, only those ions are detected whose vectorial migration to the anode is less than the cathode-directed EOF. Rapidly migrating anions escape detection. By reversing the polarity of the potential source, these fast anions can be determined but then the slow ones migrate with the EOF back into the cathode compartment. To determine the fast and slow anions in a single analysis requires the EOF to be suppressed or, better, the polarity reversed.

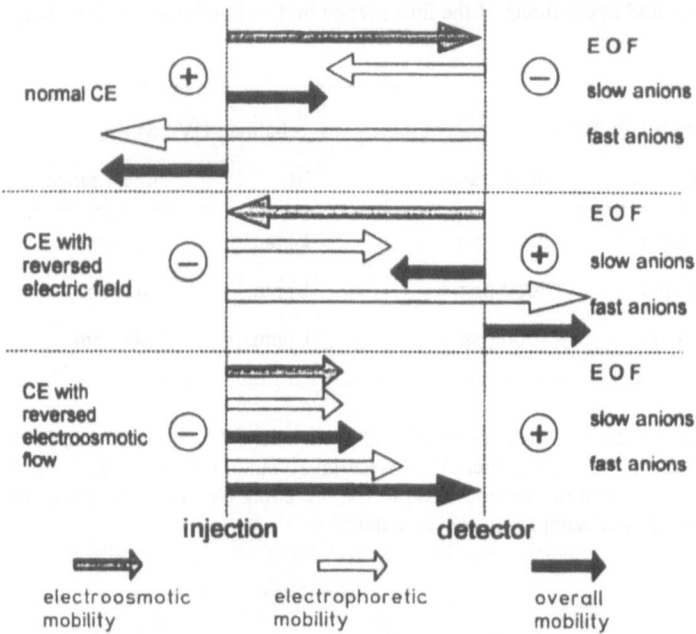

Fig. 5-16 Migration directions in anion separations by CE

Filled arrows: electrophoretic mobilities, open arrows: absolute mobilities and EOF,
Case A: normal mode, EOF toward the cathode, outlet grounded   ⎫ comigration
Case B: reversed polarity mode, EOF toward the cathode, inlet grounded ⎭
Case C: reversed polarity and anodic flow: EOF toward the anode, inlet grounded. Contramigration

The problems encountered in anion analysis by CE are presented in Fig. 5-16. In the conventional arrangement (A) with cathodic EOF, only the slow anions are detected, whose migration distance together with the magnitude of the EOF are shown in Fig. 5-17.

If the polarity of the power supply is reversed (Fig 5-16 (B)), the EOF remains cathode-directed and the fast anions migrate to the detector (now on the anode side) and are monitored there. This case is shown in Fig. 5-18. Here the anions of organic acids can no longer be detected. Fig. 5-19 shows a separation of inorganic anions by this version. Fluoride is the last ion to be detected here, phosphate would appear after 20 min as a broad peak.

If, in addition to the electric field, the EOF is also reversed, e.g., by coating the surface with a surfactant containing quaternary ammonium groups, both fast and slow anions can be separated in a single run. This was shown in Fig. 3-7 and described in Section 3.3 (Fig. 5-16 (C)). The migrations and elution sequences are illus-

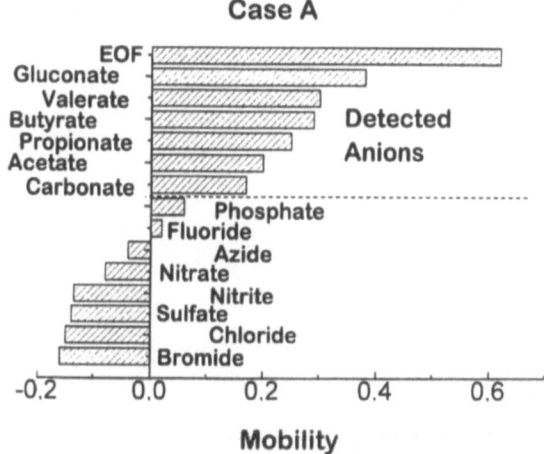

Fig. 5-17

Absolute mobilities in anion separation by CE. Case A: normal
mode, EOF toward the cathode,
outlet grounded

Separation conditions: CE instrument: Millipore Quanta 4000; capillary: 75 μm, 50/56 cm; field:
A: 600 V cm^{-1}, buffer: 5 mmol L^{-1} chromate/sulfuric acid, pH 6.8; injection: hydrostatic 4 cm, 2
s; detection: 214 nm; sample: 10 mg L^{-1}: bromide 1, chloride 2, sulfate 3, nitrite 4, nitrate 5,
azide 6, fluoride 7, phosphate 8, carbonate 9, acetate 10, propionate 11, butyrate 12, valerate 13,
D-gluconate 14, (absolute mobilities were calculated for the anions not detected)

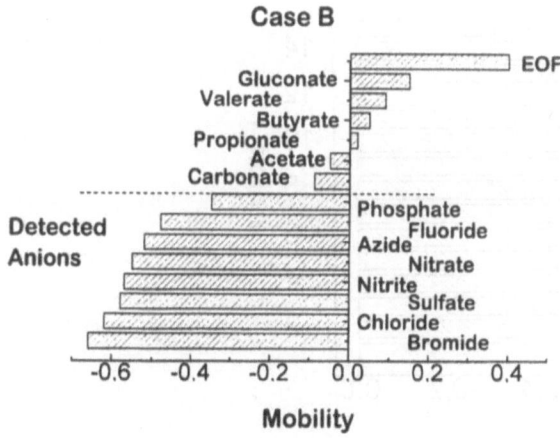

Fig. 5-18

Absolute mobilities in anion separation by CE. Case B: reversed
polarity mode, EOF toward the
cathode, inlet grounded (absolute
mobilities were calculated for the
anions not detected)

Separation conditions: CE instrument: Millipore Quanta 4000; capillary: 75 μm, 50/56 cm; field:
B: -600 V cm^{-1}; buffer 5 mmol L^{-1} chromate/sulfuric acid pH 6.8; injection: hydrostatic 4 cm, 2 s;
detection: 214 nm; sample: 10 mg L^{-1} each: bromide 1, chloride 2, sulfate 3, nitrite 4, nitrate 5,
azide 6, fluoride 7, phosphate 8, carbonate 9, acetate 10, propionate 11, butyrate 12, valerate 13,
D-gluconate 14.

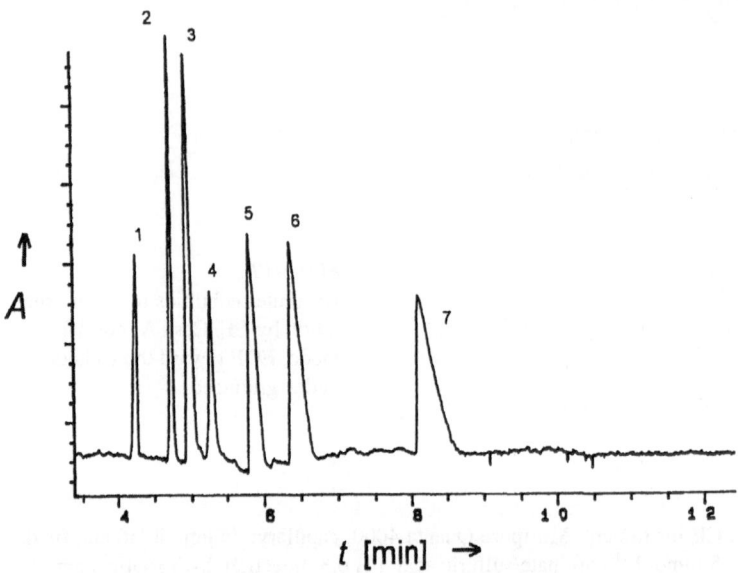

Fig. 5-19 Separation of an anion mixture according to case B

Separation conditions: CE instrument: Waters Quanta 4000; capillary 75 μm, 42/50 cm; field: -600 V cm^{-1}; detection: indirect 254 nm; injection: hydrostatic, 10 cm, 30 s; buffer; 5 mM chromate/sulfuric acid pH 6.8; sample: anion standard, anions 1 through 7 were detected

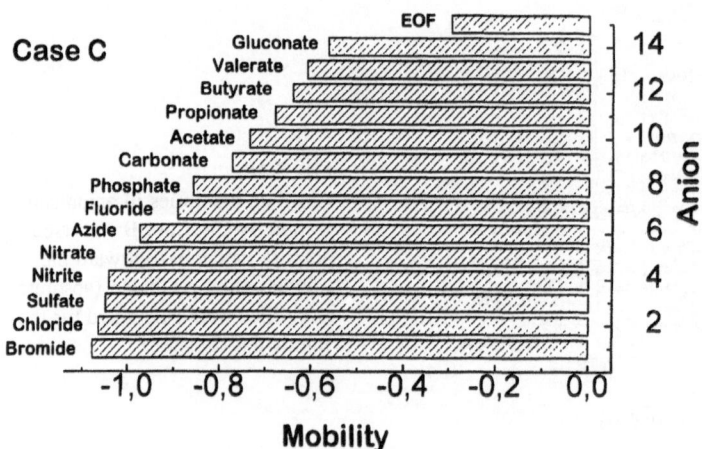

Fig. 5-20 Absolute mobilities in anion separation by CE. Case C: reversed polarity mode, EOF toward the anode, inlet grounded

Separation conditions: buffer 5 mmol L^{-1} chromate/sulfuric acid, 0.5 mmol L^{-1} CTAB, pH 6.8 (other conditions as in Fig. 5-17)

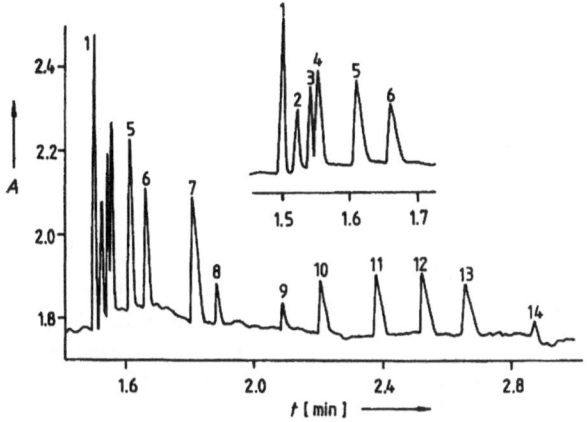

Fig. 5-21
Separation of an anion test
mixture according to case C [6]

Separation conditions: CE instrument: Waters Quanta 4000; capillary: 75 μm, 50/58 cm; data acquisition: 20 Hertz; field: -517 V cm^{-1}; detection: indirect 254 nm; injection: hydrostatic, 10 cm, 30 s; buffer: 5 mM chromate/sulfuric acid, pH 8, with 0.5 mM CTAB; sample: 10 ppm anion standard, bromide 1, chloride 2, sulfate 3, nitrite 4, nitrate 5, azide 6, fluoride 7, phosphate 8, carbonate 9, acetate 10, propionate 11, butyrate 12, valerate 13, D-gluconate 14

trated in Fig. 5-20. The corresponding separation of fast and slow anions is complete in less than 3 min, as depicted in Fig. 5-21.

With the EOF directed toward the anode, the analysis of the test mixture requires less than 2 min for the detection of fluoride and under 3 min up to gluconate which has a low intrinsic mobility. Neutral substances reach the detector after 5.5 min. The flow reversal required for this separation was effected with cetyltrimethylammonium bromide (CTAB). This cationic surfactant CTAB with its positively charged ammonium group adsorbs to the negative silanol group on the capillary wall and, at very low concentrations (less than 0.1 mmol L^{-1}), forms a layer that compensates the charge of the surface silanol groups. This treatment of the capillary stops the electroosmotic flow. If the concentration is raised to 0.2 mmol L^{-1} or more, a double layer is built up on the capillary wall. Through hydrophobic interaction, a second layer attaches itself to the first, so that the positive charge of the CTAB molecule is directed toward the inside of the capillary. This is shown in Fig. 3-7. The positive charges on the outside of this double layer is the reason for electroosmotic flow which, therefore, is no longer cathode-directed but moves to the anode. The dependence of the electroosmotic flow on the pH and CTAB concentration is shown in Fig. 5-22. One should work with CTAB concentrations of 0.5 to 1 mmol L^{-1} because at lower concentrations long equilibration times and irregularities in the EOF are observed. At higher CTAB concentrations micelles are formed which affect the selectivity as well as the efficiency of the separation. The pH dependence of the EOF, i.e., its decrease between pH 5 and 6 can be explained by the contribution of the silanol groups.

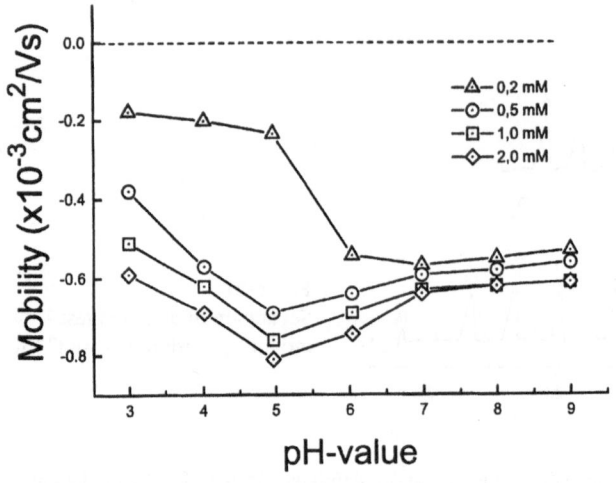

Fig. 5-22
Dependence of the electroosmotic flow on pH with different concentrations of CTAB

Separation conditions: CE instrument: Beckman P/ACE 2000; capillary 75 µm, 40/47 cm; field: -532 V cm⁻¹; detection: 214 nm; injection: pressure 3 s; buffer 10 mM phosphate; neutral marker: benzyl alcohol

In addition to CTAB, other compounds can be used, such as dodecyltrimethylammonium and tetradecyltrimethylammonium salts or hexadecylpyridinium chloride. Dodecyltrimethylammonium chloride, in particular, possesses two advantages over CTAB: its critical micellar concentration is higher and its water solubility is greater, which permits use of more concentrated stock solutions.

Chromate ion, used at concentrations between 2 and 10 mmol L⁻¹ in sulfuric acid solution, is suitable as a UV absorbing anion to be displaced by anions having average to high mobilities. The ready-to-use buffer with CTAB and chromate has low stability and must therefore be prepared fresh daily, as after a few hours a fine crystalline precipitate is formed. If one continues to work with this buffer, spikes appear in the pherogram that prevent evaluation of the results. Working with stock solutions is very time-saving. In using a 50 mmol L⁻¹ chromate solution (with 0.2 mmol L⁻¹ sulfuric acid) and a 50 mmol L⁻¹ CTAB solution, just prior to use 10% (v/v) of the chromate solution must be pipetted into a volumetric flask, filled to about 2/3 with water, 1% (v/v) of the CTAB stock solution added, and the solution is made up to the mark with water. The stock solutions are stable for several months. The solubility of a 50 mM CTAB solution is exceeded at 18°C, but above 25°C a clear solution is formed again rapidly. Only with strict adherence to this sequence of steps is precipitation avoided in the preparation of the buffer. A good description of the method and a listing of more than 50 ions that were measured with indirect UV detection are given in [36].

Because of the extremely high efficiency (300,000 - 500,000 plates per meter) and the short analysis times, problems stemming from the rate of data acquisition cannot be ruled out. Peak widths of less than one second reach the limits of data acquisition of most CE instruments: The 20 points per peak required for a good integration are not attained with a data acquisition rate of 20 Hertz.

A summary of buffer additives that can be used to effect flow reversal or a reduction in the EOF is presented in Table 5-4. Quaternary ammonium compounds and polyamines always lead to flow reversal, whereas simple amines or zwitterionic compounds strongly reduce the EOF. These substances neutralize the negative surface charges, but are not able to build a double layer and thereby to generate an excess of positive charge on the capillary surface. If instead of flow reversal only a retardation of the EOF is brought about, the separation of anions requires 10 to 15 min. This, however, produces no advantages for anion separation.

As has already been explained for cation separation, it is essential that the mobilities of the solutes and the background electrolyte be similar. Chromate is optimal for the analysis of inorganic ions because of its high mobility. Organic anions require background electrolytes with lower mobilities. Table 5-5 provides a compilation of some useful background electrolytes and their mobilities, pH ranges, and optimal detection ranges.

Table 5-4 Overview of means of modifying the EOF

Buffer additive or coating	useful concentrations	Effect	Problems
CTAB	0.2 mM - 1 mM	flow reversal	micelle formation, solubility to 50 mM
DTAB	0.2 mM - 2 mM	flow reversal	micelle formation,
HDP	0.2 mM - 1.5 mM	flow reversal	micelle formation, UV absorption
SB 14	1 mM -10 mM	very low EOF	
Spermine	2 mM - 10 mM	very low EOF	ion-pair formation
Diethylenetriamine	1 mM - 5 mM	very low EOF	
Polyethylenamine	0.5 mM-2 mM	flow reversal	
SAX coating	–	flow reversal	low stability

CTAB: Cetyltrimethylammonium bromide, DTAB: dodecyltrimethylammonium bromide, HDP: hexadecylpyridinium chloride, SB 14: tetradecyldimethylammonium propanesulfonic acid

Table 5-5 Overview of buffer substances for indirect UV detection of anions

Buffer ion	Mobility [cm^{2}kV^{-1}s^{-1}]	pH range	UV range [nm]	Problems
Chromate	~ 0.8	6-10	<220 und 250-280	carcinogenic
Molybdate	~ 0.7		<290	isopoly acids above pH6
Sorbic acid	0.24	>3	225-275	
N-Ac-tryptophane	0.23	>4,5	<265	
Tryptophane	0.23	>8,5	<265	
3-Indoleacetate	0.20	>5	214-275	
3,5-dinitrobenzoic acid	0.22	>2	<260	
Trinitrobenzenesulfonic acid	0.20	>2	220-275	available only as a solution
Dipicolinic acid	0.23	>5	<220 und 245-265	complexation with calcium
Naphthol-1-sulfonic acid	0.16	>2	200-225	low solubility in water
Gallic acid	0.26	>3	<230	unstable (oxidation)

-The mobilities were determined at pH 9.0 in 10 mM phosphate buffer

The anion analysis of a water sample of Lake Baikal represents another application of the chromate buffer system. The separation is presented in Fig. 5-23. The ions present at higher concentrations (chloride, sulfate, and carbonate) were quantified by standard addition, whereas nitrate and phosphate were determined by comparison with external standards because of their low concentrations. The advantages of CE in comparison to IEC lie not only in the substantially shorter analysis times, but also in the ability to determine bicarbonate in addition to the other anions.

The separation of free fatty acids is presented as an example of the analysis of organic acids. Fig. 5-24 shows the analysis with 3,5-dinitrobenzoic acid as background electrolyte. To improve the solubility of the higher fatty acids, 30% acetone was added to the buffer. The formation of the double layer for flow reversal is not impaired by the addition of the acetone, which is not the case for methanol and 2-propanol. Only in acetone concentrations exceeding 50% does the double layer break down.

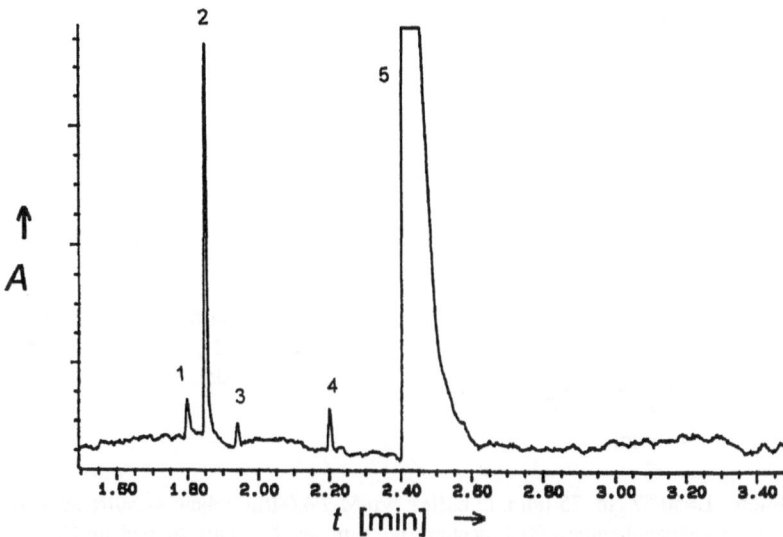

Fig. 5-23 Determination of anions in surface water (Lake Baikal)

Separation conditions: CE instrument: Waters Quanta 4000; data acquisition rate: 20 Hertz; capillary: 75 µm, 50/58 cm; field: -431 V cm^{-1}; detection: indirect, 254 nm; injection: hydrostatic, 10 cm, 30 s; buffer: 5 mmol L^{-1} chromate/sulfuric acid pH 8.0, 0.5 mmol L^{-1} CTAB; peak assignment: chloride 1 (0.5 ppm), sulfate 2 (12.4 ppm), nitrate 3 (<0.5 ppm), phosphate 4 (<0.5 ppm), bicarbonate 5 (81.1 ppm)

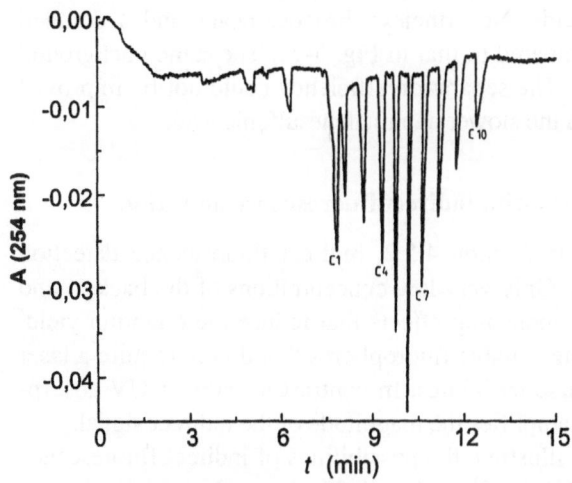

Fig. 5-24
Analysis of free fatty acids
with reversed EOF

Separation conditions: L=50/57 cm, 75 µm i.d., buffer: 5 mM 3,5-dinitrobenzoic acid pH 9.0; 0.5 mM CTAB, 40 % (v/v) acetone; samples: free fatty acids C_1 - C_{10} (25 ppm each), potential: 25 kV, reversed polarity, detection at 214 nm

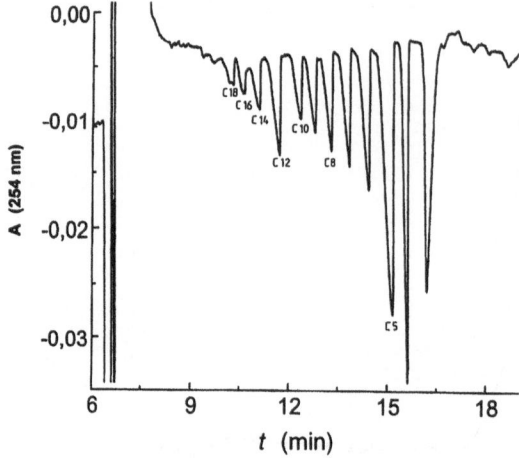

Fig. 5-25
Analysis of free fatty acids in
the normal EOF mode (toward
the cathode)

Separation conditions: L=50/57 cm, 75 µm i.d.; buffer: 5 mM 2,4,6-trinitrobenzenesulfonic acid, pH 10.7 (adjusted with triethanolamine); 60 % acetonitrile; samples: free fatty acids from C_3 - C_{10}, C_{14}, C_{16}, C_{18} (25 ppm each), C_{12} (50 ppm); potential: 25 kV, reversed polarity; detection at 254 nm

The migration of the fatty acids is slow enough for them to be analyzed without flow reversal. However, the migration sequence is then reversed, with the long chain fatty acids being detected first, as can be seen in Fig. 5-25. The other separation conditions are identical. A higher acetone content may be employed to enhance the solubility of the longer chain fatty acids. Nevertheless, broader peaks and a doubled analysis time were obtained as compared to that in Fig. 5-24. The same background electrolyte was used in both cases. The separation efficiency could not be improved from that depicted in Fig. 5-25 with the slower naphthalenesulfonic acid.

5.2.4 Analysis of cations and anions with indirect fluorescence detection

As has already been pointed out in Section 4.5.2, indirect fluorescence detection presents some additional problems. Only very low concentrations of the background electrolytes may be used to avoid quenching effects that reduce the quantum yield. There are also problems in choosing suitable fluorophores that do not require a laser for excitation and exhibit appropriate mobilities. In contrast to indirect UV absorption, self-absorption by the analyte amplifies the magnitude of the indirect signal.

Two examples are presented to illustrate the possibilities of indirect fluorescence detection compared to indirect UV detection. Fig. 5-26 shows the cation analysis with 9-aminoacridine as background electrolyte. A modified HPLC fluorescence detector was used for these studies. The detection sensitivity for the alkali ions was

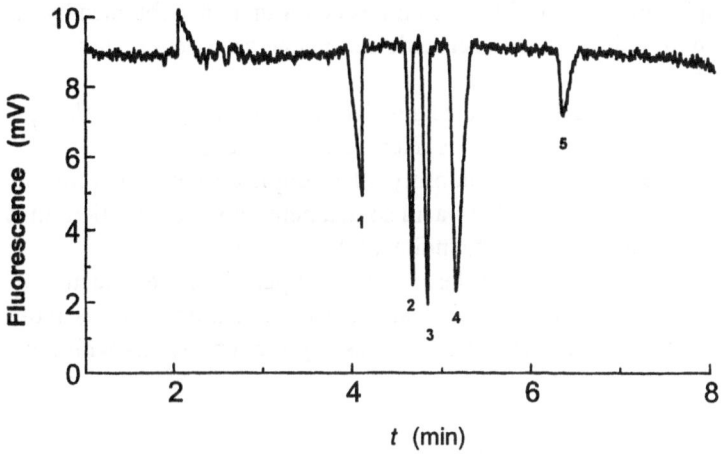

Fig. 5-26 Indirect fluorescence detection of cations

Separation conditions: L=50/62 cm, detection window enlarged to 105 μm, 1.5 mM 9-aminoacri-dine/citric acid pH 3.0, potential: 20 kV, detection: modified HPLC fluorescence detector (HP 1046 A), excitation wavelength: 247 nm, emission wavelength: 440 nm. sample: potassium: 78 ppb (1), sodium: 46 ppb (2), magnesium: 48 ppb (3), calcium: 80 ppb (4), lithium: 13 ppb (5)

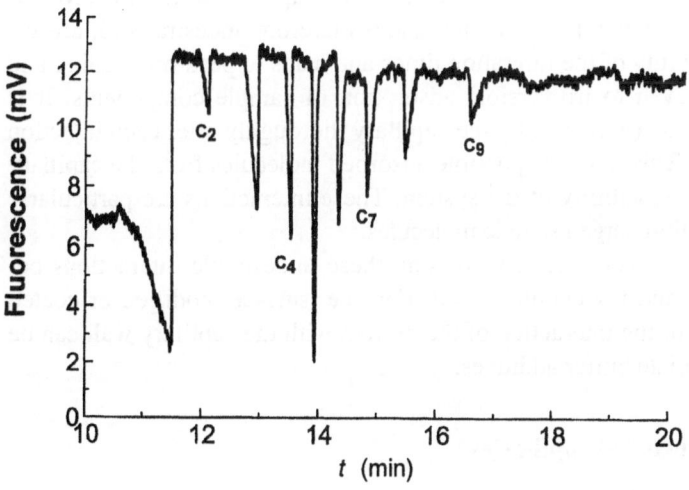

Fig. 5-27 Indirect fluorescence detection of free fatty acids

Separation conditions: 2.5 mM salicylate, 0.3 mM CTAB, 10 % acetone pH 11, detection: modi-fied HPLC fluorescence detector (HP 1046 A), excitation wavelength: 229 nm, emission wave-length 415 nm; sample: free fatty acids C_2 - C_9 (250 ppb each)

found to be about 5 ppb and was thus 50 to 100 times better than that obtained by in-direct UV detection. With indirect fluorescence detection one can operate with CE in the sensitivity range of suppressed ion IEC.

Sodium salicylate is well suited as background electrolyte for the indirect fluores-cence detection of anions, as Fig. 5-27 shows. Here, too, flow reversal at pH 11 was used and acetone was added to improve solubility. The sample concentration for this pherogram amounts to 250 ppb of each fatty acid so that here, too, the detection limit could be reduced about 50-fold compared to indirect UV detection.

Indirect fluorescence detection without laser excitation permits trace detection of cations and anions by CE. The manipulation and the optimization requires more knowledge about the optimization steps than is necessary for separations with indi-rect UV detection.

5.3 Capillary Zone Electrophoresis of Proteins

In the CZE of proteins, strong interactions between the analyte molecules and the capillary wall may appear that are predominantly electrostatic in nature between the negatively charged silanol groups and the positively charged functionalities of the analyte. Additional nonspecific interactions such as hydrogen bonding or van der Waals forces (protein-protein interaction) may also play a role.

The adsorption of molecules such as proteins on the capillary wall may have ad-verse effects on the electrophoretic separation and is therefore undesirable. It always degrades the reproducibility of the migration times and leads to peak broadening and peak asymmetry, and even to irreversible adsorption of sample components. It is therefore essential to flush (with NaOH) the capillary thoroughly after each injection of biological matrices. This removes possible adsorbed molecules from the capillary wall and raises the reproducibility of the system. These interactions are particularly strong for positively multi-charged sample molecules.

There are several approaches for suppressing these undesirable interactions be-tween the analyte ions and the capillary wall. For one, surface modified or coated capillaries can be used or the interaction of the protein with the capillary wall can be suppressed with appropriate buffer additives.

5.3.1 Separations in uncoated capillaries

5.3.1.1 Selection of the pH

Proteins, being amphoteric substances, are positively charged when the buffer pH is greater than their pI values and negatively charged at lower pHs. Therefore, most protein molecules have only a small net charge since the positive and negative

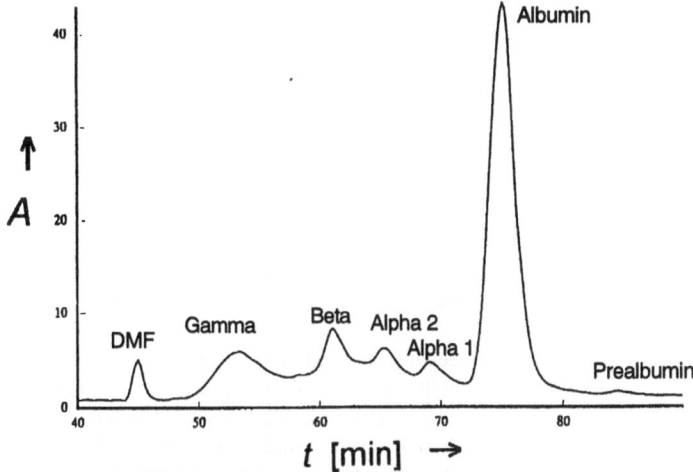

Fig. 5-28 Separation of serum proteins at pH 10 in an untreated capillary [41]

Separation conditions: capillary: 25 cm, 25 μm i.d., potential: 25 kV, buffer: phosphate pH 10; sample: human blood serum

charges tend to compensate each other. This net charge, however, governs the direction of migration and the velocity, whereas more important for wall adsorption is the absolute number of positive charges in the sample molecule that can interact with the negatively charged silanol groups. Above their pI, proteins have the same nature of charge (net charge) as the capillary wall and, theoretically, are repelled from it during separation. Since many proteins have pI in the neutral to slightly acidic range, the electrophoresis buffer must have a distinctly alkaline pH (in the range of 9 - 11). However, this approach to optimization reaches its limits in the case of strongly basic proteins (pI>10). Furthermore, the high conductance of such buffers and the denaturation phenomena that may appear under these drastic conditions may have negative consequences. Fig. 5-28 shows a protein separation in an unmodified capillary at pH 10. The serum proteins are separated into fractions comparable to those in classical gel electrophoresis. But quantification in CE is more accurate. Due to the short analysis times in CE, sample throughput is comparable to that of classical gel electrophoresis, even though the separations must be carried out sequentially in CE, whereas they can be performed in parallel by the classical method.

In the strongly acid pH range (pH<2), protein adsorption can be reduced by the protonation of the silanol groups and thus the elimination of the negatively charged surface. Problems that arise are the very low EOF and the possible denaturation of the proteins.

The use of extreme pH values for the analysis of biopolymers limits the selectivity of the system because the charge differences of the analytes are thereby drastically re-

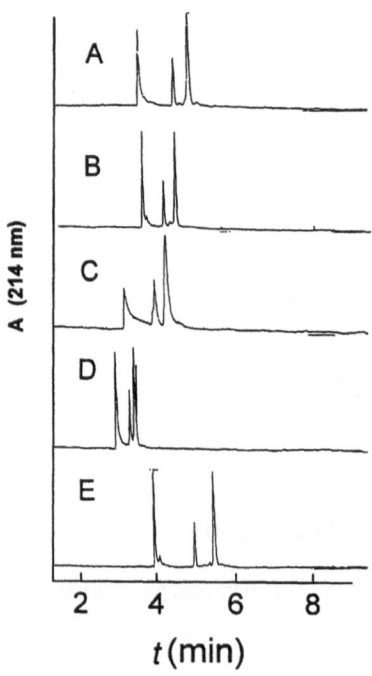

A (214 nm)

2   4   6   8

t (min)

Fig. 5-29
Separation of protein standards in different buffers

Separation conditions: instrument: laboratory-built with a fast scanning detector; capillary: 50 μm, 37/44 cm; field: 341 V cm^{-1}; injection: hydrodynamic 15 cm, 2 s; detection: 214 nm; phosphate buffer; sample: cytochrome C, lysozyme, ribonuclease A. Buffer: a) 50 mM pH 4.3; b) 50 mM pH 2.9; c) 20 mM pH 2.3; d) 20 mM pH 2.3; e) 100 mM pH 4.2

duced. Moreover, it is advantageous to maintain the pH as a freely selectable parameter for the optimization. The migration velocity of charged samples is determined by the degree of ionization, which can be readily altered via the pH value.

The best results are achieved at high buffer concentrations and an acid pH range just below 3. Fig. 5-29 shows some buffers that were tested in the pH range 2.3 to 4.7 for the separation of basic proteins. By raising the pH from 2.3 to 4.7, a distinct reduction in the peak efficiency was noticeable at a buffer concentration of 25 mmol L^{-1}, whereas in raising it from 50 mM (pH 4.3) to 100 mM (pH 4.2) a significant increase in the peak efficiency could be observed. By decreasing the pH to 2.9, even at these high buffer concentrations an additional increase in the efficiency was found, although the current rose considerably due to the extremely high ionic mobility of protons (acidic pH). Further optimization to still lower pH values could only be achieved by simultaneous dilution of the buffer, which is again accompanied by a reduction in the peak efficiency. In addition, the very acidic pH region of 2.3 led to a lowering of the selectivity.

5.3.1.2 Addition of salts to the buffer

The protein-wall interaction can be attributed primarily to a cation exchange mechanism. This is plausible insofar as even with a negative overall charge on the molecule (i.e., especially in basic pH), there are still cationic groups (e.g., arginine residues) present in the polypeptide chain. As in ion exchange chromatography, the addition of higher concentrations of alkali salts (e.g., potassium sulfate) to the electrophoresis buffer furnishes competing ions for the Coulombic attraction and thereby drastically reduces the protein-wall interaction.

Using this concept, efficiencies of 50 - 100,000 plates per meter were attained for standard proteins over a broad pI range (pI 5 - 11). Disadvantages here stem from the comparatively very high conductances of the electrophoresis buffers (efficient cooling is essential) that necessitate the use of lower fields (5kV) and long capillaries of small inner diameter (25 μm) (diminished detection sensitivity). Moreover, the high ionic strengths reduce both the electroosmotic flow and the ζ-potential of the samples.

Zwitterionic molecules (inner salts) can serve as buffer substances that possess a high buffer capacity but do not contribute appreciably to the overall conductance of the system. Zwitter ions minimize protein adsorption by interacting with the proteins and capillary wall. By adding, e.g., amino sulfonates, amino sulfates, and, recently, phosphonium sulfonates in very high concentrations to phosphate buffers, both basic and acidic proteins can be resolved. An overview of the frequently used separation systems for proteins in uncoated capillaries is shown in Table 5-6.

5.3.1.3 Use of buffer additives for the separation of proteins

The simplest way to modify the surface of quartz capillaries is to add a component to the buffer that is preferentially adsorbed by the surface silanol groups. The formation of this layer affects the EOF and also reduces adsorption of the analyte through hydrophobic interaction or electrostatic repulsion.

Table 5-6 Protein separations with buffer additives with uncoated capillaries

Buffer	C [mM]	pH	pI	N $[\times 10^3 \ m^{-1}]$	Lit.
Tricine/KCl	10/20	8.3	5 - 7	500	[42]
CHES/K$_2$SO$_4$/EDTA	100/250/1	9.0	5 - 11	100	[43]
Phosphate/betaine/K$_2$SO$_4$	40/2000/100	6.7	9 - 11	300	[44]
Phosphate/sultone	100/1000	7.0	5 - 11	300	[45]
1,3-Diaminopropane/K$_2$SO$_4$	40/20	3.5	9 - 11	300	[46]
Phosphate/BRIJ 35	10/10 ppm	7.0	3 - 11	100	[47]
Phosphate/FC 134	50/100 ppm	7.0	7 - 11	300	[48]
Borate/SDS	100/0,5 %	10.0	4 - 6	–	[49]

First of all, the addition of *surfactants* to the buffer can improve the efficiency of protein separations. The interactions between various surfactants and analytes can differ considerably. For one, the adsorption of the detergent onto the biomolecule could certainly be discussed. The consequence of this is improved solubility and increased hydrophilicity of the sample. The adsorption of surfactants onto the capillary wall also plays a large role in increasing the hydrophilicity of the surface and possible blocking of the adsorption sites.

With *cationic detergents* (e.g., cetyltrimethylammonium bromide or "FC 134" , a perfluorinated C8-sulfonamide with a quaternary ammonium group (see Table 5-6)), a double layer is formed in which the positive charges point toward the inside of the capillary, as has already been discussed for the reversal of the EOF in anion analysis. The positive surface charge and the consequent electrostatic repulsion impedes the adsorption of cationic proteins. High plate numbers and symmetrical peaks are obtained with basic proteins. Care must be taken not to exceed the critical micellar concentration of the detergent, as this leads to a completely different migration behavior (*cf.* MEKC)

Uncharged surfactants (niotensides) such as the nonionic surfactant BRIJ 35 (see Table 5-6) can also be used to reduce protein-wall interactions. To strengthen the sorption of these detergents on the capillary surface and to completely suppress the ionic interaction with silanol groups, a "thick" coating on the wall of conventional C18 is also often used.

It should also be possible, in principle, to use in CZE the *anionic surfactant* sodium dodecyl sulfate (SDS) that has especially proven itself in classical electrophoresis and in LC. Systematic investigations with acidic proteins have shown, however, that the addition of SDS to the sample and to the buffer in the ppt or per cent range yield very unclear results. Overall, neither efficiency nor selectivity improves, but sometimes rather dramatic deterioration occurs. These results can perhaps be explained by the well known denaturation effect of SDS on proteins. Besides, the uniform negative charge of biomolecules resulting from adsorption of the surfactant certainly plays a significant role in the observed loss in selectivity. The original charge structure of the protein can be completely lost or obscured through interaction with SDS, so that the separation in an electric field is based on the charges developed due to adsorption of detergent ions. This effect is used in SDS-PAGE for the estimation of the molecular weights of proteins. This technique can also be carried out in capillaries and will be described later in the chapter on capillary gel electrophoresis.

In summarizing the use of detergents for protein separations by CE, the most promising approach involves the nonionic surfactants as a dynamic coating because they do not appreciably alter the charge structure of the samples. However, this method often involves a combination of chemical surface modification and dynamic coating, so that the advantage of ready manipulation and accessibility afforded by the use of buffer additives is lost.

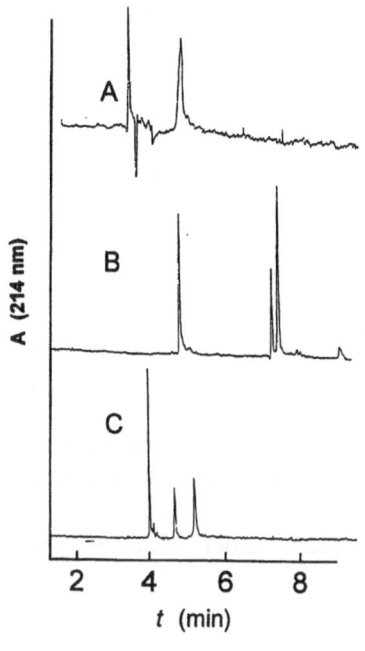

Fig. 5-30
Separation of protein standards in a DAP-buffer system

Separation conditions: capillary: 75 μm, 37/44 cm; field: 272 V cm^{-1}; detection: 214 nm; phosphate buffer with 100 mM DAP pH 3.0; sample: cytochrome C, lysozyme, ribonuclease A,
a: 50 mM phosphate, 100 mM DAP, pH 10.7
b: 50 mM phosphate, 100 mM DAP, pH 6.5
c: 50 mM phosphate, 100 mM DAP, pH 3.0

5.3.1.4 Dynamic coating of capillaries

The undesirable adsorption of proteins on the capillary surface can also be reduced by adding the *lower polyamines* (e.g., DAP, 1,3-diaminopropane) to the buffer. High efficiencies are obtained, but due to the high conductance of the buffer the field strength must be kept low or capillaries with 25 to 50 μm i.d. must be used. The effectiveness of the buffer additive can presumably be attributed in part to its irreversible adsorption on the capillary surface. This is indicated by the reduced adsorption of the analyte molecules and the greatly diminished electroosmosis. This is exemplified by separation of standard proteins in Fig. 5-30. At high pH values (curve A) there is hardly any adsorption of DAP on the surface and a useful separation is no longer obtained. High efficiencies are obtained in both acid and neutral pH ranges.

5.3.2 Protein separations with surface-modified capillaries

The EOF can also be controlled by chemical modification of the capillary surface. This results in reduced adsorption of the sample components on the capillary surface and improved reproducibility of analyses. Various options that were developed for the modification of silica in HPLC and for coating capillaries in GC are available for the chemical coating of capillaries.

5.3.2.1 Coatings for capillary electrophoresis

In principle, two different techniques for surface coating can be distinguished. Initially, the traditional silane chemistry was used primarily, in which mono-, di-, or trifunctional silanes were attached to the silanol groups of the capillary wall via a siloxane bond. The functional groups introduced via the bonded silanes can, if necessary, be converted in a second step to the final coating.

A compilation of such conventionally coated capillaries is given in Table 5-7. The limitation of this technique lies in the poor hydrolytic stability of siloxane bonds at pH values above about 8, as has been long known from liquid chromatography.

A way to circumvent these disadvantages is to use polymeric wall coatings, which represent the second large group of modification techniques. Again, two different approaches can be distinguished:

-   The surface is pretreated by classical silane chemistry; so-called anchor groups are introduced. These can then be copolymerized with the appropriate monomers or oligomers in a second step.
-   Prepolymers are first adsorbed on the corresponding carrier, which are then copolymerized *in situ* and cross linked.

Table 5-7 Overview of capillary coating processes

Conventional coatings			
Functionality	pH range*	Applications	Literature
Trimethylsilyl-	7	small molecules	[50]
Trimethylsilyl-	9	MECC	[51]
RP-C8	9	proteins	[52]
RP-C18	9	proteins	[52, 53]
Polyethylene glycol-	3 to 5	proteins	[54]
Diol-	3 to 5	proteins	[50, 55]
Glyceroglycidoxypropyl-	5	proteins	[56]
Maltose	3 to 7	proteins	[55]
Arylpentafluoro-	7	proteins	[57]
a-Lactalbumin	8	proteins	[58]

*stability range given in the literature or the highest pH used in measurements

Polymer coated surfaces exhibit improved pH stability in the alkaline range up to about pH 9. The disadvantage of polymer coatings is their strong hydrophobicity (adsorption of proteins!), so that such phases are frequently used in the presence of nonionic or zwitterionic detergents. A compilation of the polymer coatings that have been used is given in Table 5-8.

Since the CZE of proteins represents an important separation technique whose success depends very strongly on the type of capillary used and its surface modification, the individual coating techniques, their characterization, and their advantages and disadvantages will be discussed in detail. Because understanding the chemistry of surface modification is important for gaining insight into possible protein interactions with the surface, the preparation methods of the various types of coatings will be described in detail.

The characterization of capillaries before and after modification is extremely difficult. In GC and LC numerous standard tests and test mixtures are available for characterizing stationary phases. Furthermore, especially in LC, there are many independent physicochemical methods such as elemental (CHN) analysis, FTIR or NMR to obtain definitive information on surface modification. However, the short tubes (L_{tot} < 1 m) of extremely small inner diameters (d < 100 μm) and the corresponding

Table 5-8 Overview of capillary coating processes

| Functionality | Polymeric coatings | | |
	pH range*	Applications	Literature
Linear polyacrylamide	2 to 8	proteins	[59-61]
	(up to 10)	IEF	[62, 63]
		DNA-fragments	[64]
Polyethyleneimine	3 to 11	proteins	[65]
1-Vinyl-2-pyrrolidone	2 to 6	proteins	[56]
Poly(methylglutamate)	7	proteins	[66]
Polymethylsiloxane (OV 1)	7	MEKC	[67, 68]
Polyethylene glycol	7	MEKC	[67, 68]
Polyvinyl alcohol	8	proteins	[72]
Polysiloxane/ β-cyclodextrin	7	chiral separations	[69, 70]

*stability range given in the literature or the highest pH used in measurements

inner surface area of only a few cm^2 that are used in CE have proven themselves to be inaccessible to instrumental physicochemical methods.

Therefore, the change in the electroosmotic flow caused by the coating is viewed as characteristic of it. Moreover, since most coatings were developed to suppress protein-wall interactions, it is not surprising that the separations of protein standard mixtures serve as an index of the effectiveness of the surface modifications achieved.

Polar or hydrophobic coatings with good coverage can also be characterized by GC measurements. The retention behavior of n-nonane as a standard substance is utilized to determine the film thickness. The retention behavior of various polar substances serves to characterize the silanol groups still present.

From the many years of experience in the preparation of coated quartz capillaries for GC, it is known that pretreatment of the material prior to actual chemical reaction is very beneficial. The capillaries used in CE involve either "not deactivated" GC capillaries or byproducts from fiber optics production. The condition of the various raw materials with respect to "surface roughness" and contamination with metal ions differs widely. As has already been discussed in Sections 3.3 and 4.2, various materials differ in the magnitude of the EOF in acidic and basic media. Therefore, most methods described in the literature start with a treatment with alkalis or acids (leaching) to enhance the wettability of the surface. Moreover, the surface is also attacked chemically and new silanol groups are formed. This is beneficial as the additional SiOH groups can serve as further bonding sites for the chemical modification and thereby aid in improving the yield of the chemical reaction with the capillary wall. This step is usually followed by washing with distilled water until neutral and a drying step in a gas stream at elevated temperatures (80 - 200°C). Not all authors point out the absolute necessity of pretreatment for a successful and reproducible coating.

Two basic methods are known from GC for modifying the capillary surface. The first process involves *dynamic coating* [71]. Here a plug of liquid consisting of an appropriate solvent and the dissolved stationary phase is forced through the capillary column by means of a gas stream. Important conditions include that the solvent wet the carrier wall well, i.e., that the solutions have a low surface tension and, above all, are dust-free. Dichloromethane and pentane are the main solvents used. The thickness and homogeneity of the resulting coating is governed by the solution concentration. The film thicknesses attainable by this method lie between 0.2 and 1 μm for a 10% solution. The coating is followed by drying in an inert gas stream and possible polymerization or crosslinking of the film.

In the *static method* [71] the capillary is also filled with a dilute solution of the stationary phase. After fusing off one end, the solvent is evaporated off through the open end by heating and/or reduced pressure. This technique poses substantially higher instrumental and experimental requirements. Prerequisites for a homogeneous coating most importantly include extreme temperature constancy and careful degassing of the solution. Even though both techniques have proven themselves thoroughly in GC, they have thus far been seldom used for CE. Mostly, the coating techniques of HPLC are used, in which, e.g., the capillary is filled with a solution of a silane and the reaction is allowed to proceed by standing or is accelerated by heating. This is accompanied by poorer controllability and reproducibility of the coating. Moreover, important information about the modified surface (film thickness, see above) is lost.

5.3.3 Overview of important chemical coatings for protein separation

5.3.3.1 Conventional coatings

For coating with *alkylsilanes*, mono-, di-, and trifunctional silanes are used, as is well known from GC and HPLC. The capillary is filled with a solution of a reactive silane and the reaction is initiated by raising the temperature. The resulting surface possesses more or less hydrophobic properties (reversed phase) depending on the chain length of the bonded aliphatic residue. Owing to their lower reactivity, alkoxysilanes can also be bonded from aqueous solution.

The reaction of the surface silanol groups with neutral (uncharged) alkylsilanes leads to a reduced electroosmotic flow. Fig. 5-31 compares the dependence of electroosmosis on the buffer pH for capillaries coated with C8 and C18 brushes and the starting material. It can be seen that the flow is reduced to about 40% of its original value and the inflection point of the curve cannot be ascertained unambiguously. The lower flow for the C8 capillary can probably be explained by the better surface coverage as C8 brushes have lesser steric requirements than C18 brushes – as is known from the preparation of reversed phases - and a higher degree of coverage is attained.

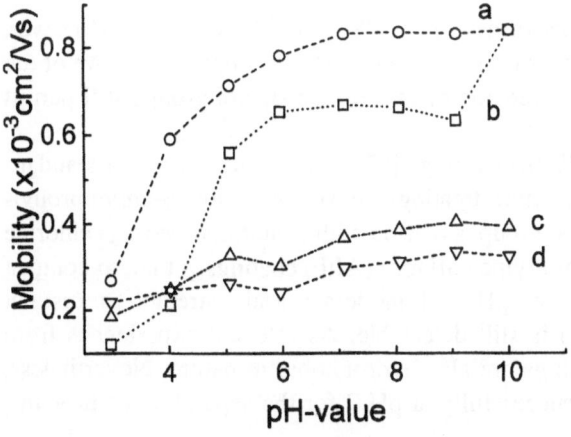

Fig. 5-31
EOF/pH characteristics of various coatings (brush types) [52]

a: uncoated capillary
b: polar coating
c: strongly hydrophobic coating (C-18)
d: weakly hydrophobic coating (C-8)

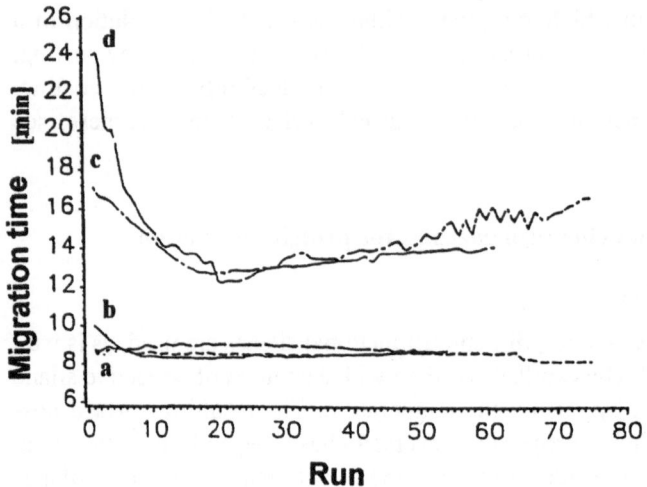

Fig. 5-32 Stability of coatings (brush types) [52]

a: uncoated capillary
b: polar coating
c: strongly hydrophobic coating (C-18)
d: weakly hydrophobic coating (C-8)

It is known from chromatographic applications of reversed phases that alkylsilane coatings are hydrolytically stable only up to a pH of about 7. Fig. 5-32 presents the course of the migration times of a neutral marker over 80 runs for the above capillaries. The modified capillaries show a distinct drop-off during the first 20 runs. Possible explanations for this are either a loss of unbonded material or partial hydrolysis of the brushes from the surface. Thereafter, at least for the C8 capillary, a stable condition is attained.

Protein separations can be performed with C8 and C18 coatings. However, alkylsilane coatings possess two major disadvantages. The hydrophobic nature of the alkyl brushes and the incomplete screening of the surface silanol groups still permit interactions with proteins.

The preparation of *arylpentafluoro coatings* [57] follows the described standard methods as a two-step reaction. After treating the surface with γ-aminopropyl-trimethoxysilane, the bonded amino group is reacted with pentafluorobenzoyl chloride in dry toluene. A special feature of arylpentafluoro (AFP) coatings is that, in contrast to many other coatings, at neutral pH and moderate ionic strength a distinct electroosmotic flow (0.5 mm s^{-1}) is still detectable. As relevant experiences from chromatography show, AFP coatings are also hydrophobic in nature. Nevertheless, AFP capillaries have been used successfully at pH 7 for the separation of proteins.

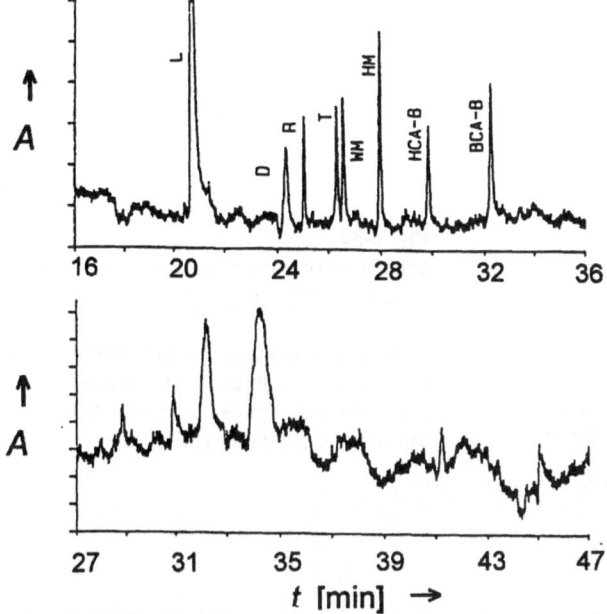

Fig. 5-33 Model separation of proteins with an AFP-coating (upper) and with a fused silica capillary (lower) [57]

Separation conditions: buffer: 200 mM phosphate, 100 mM KCl pH 7, solutes: L lysozyme, D DMSO, R RNA (bovine), T tripsinogen, WM myoglobin (whale), HM myoglobin (equine), HCA-B carboxyanhydrase B (human), BCA-B carboxyanhydrase (bovine)

For a test mixture of proteins that cover a pI range of 6.9 to 11, efficiencies of several hundred thousand theoretical plates per meter were found (Fig. 5-33). The contribution of electroosmosis to the transport of proteins permits both cationic and anionic proteins to be separated in a single run. Under the conditions used, no useful separation is obtained in the uncoated capillary (lower electropherogram in Fig. 5-33).

In LC, for protein separations principally hydrophilic phases are used, e.g., diol-, polyethylene glycol- (PEG), or polysaccharide-coated silica.

For coating quartz capillaries with *hydrophilic polyether groups* a two-step reaction is always used, in which, for example, a glycidsilane is bonded. This can then be either hydrolyzed to a diol or reacted further with glycerol or polyethylene glycol. The binding of carbohydrates is effected by first attaching an amino function to which then e.g., maltose can be bonded as a Schiff's base and the double bond then reduced by cyanoborohydride. Diol and polyethylene glycol coatings exhibit, as expected, strongly reduced electroosmosis (diol: $\mu_{EOF} = 0.1$ cm^2 kV^{-1} s^{-1} in 50 mM phosphate, pH 6). The operating range for these capillaries is given as pH 3 - 5 and their long-

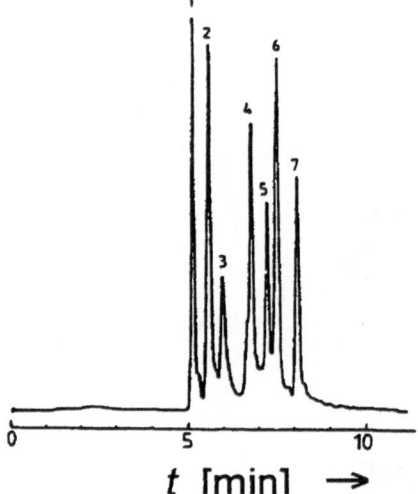

Fig. 5-34
Separation of protein standards with a PEG coated capillary [54]

Separation conditions: buffer: 30 mM phosphate, pH 3.8; solutes: 1 cytochrome C, 2 lysozyme, 3 myoglobin, 4 trypsin, 5 RNA, 6 tripsinogen, 7 chymotrypsinogen

term stability under these conditions is several months. Those coated with maltose can be used in the pH range of 3 – 7. Since in the reaction of the aminosilane with maltose not all amino functions are converted, the surface is positively charged at low pH values. This leads to a reversal in the direction of flow of the EOF in acid solution. Moreover, the capillary can only be stored with the addition of a preservative to prevent attack from microorganisms. An efficient separation of standard proteins with a PEG capillary at pH 3.8 is shown in Fig. 5-34 [54]. Even with these hydrophilic coatings, the capillary surface or the "foundation" introduced in the first synthetic step is often insufficiently covered by the polar final layer to exclude appreciable interactions between the sample and silanol or amino groups.

Like carbohydrates, *proteins* (e.g., α-lactalbumin) can also be bonded onto the capillary wall [58]. For this, a capillary modified with amino groups is treated with glutaric dialdehyde in order to introduce free aldehyde groups as protein binding sites. These finally react with the protein that is pumped through in aqueous solution and is thus chemically anchored. The protein coatings exhibit interesting EOF/pH characteristics. Since proteins possess amphoteric properties, they are present in the anionic form above their pI and in the cationic form below it. If the buffer pH corresponds to the pI value, the proteins are uncharged. Consequently, electroosmosis should cease at this point or reverse flow direction when this point is exceeded. In principle, the magnitude and direction of the electroosmotic flow could be controlled through the choice of an appropriate protein or amphoteric molecule. Despite these interesting properties, these types of molecules have found little application so far.

5.3.3.2 Polymeric coatings

The coating with linear polyacrylamide [59-61] is one of the most frequently used coating methods. One of the reasons for that may be that cross-linked polyacrylamide gels have proven their excellence as carrier and separation matrices for the electrophoretic separation of proteins in the flat-bed techniques such as DISK electrophoresis, SDS-PAGE, or isoelectric focusing. Apparently because of their hydrophilic nature, such gels exhibit only weak interactions with biomolecules. In CE, coating of the quartz surface with acrylamide is also possible. For this, in a two-step reaction the capillary is initially treated with γ-methacryloxypropyltri-methoxysilane in dilute acetic acid solution. The olefinic groups thus attached are then copolymerized in the second step with acrylamide in aqueous solution. Ammonium persulfate is used as radical initiator and N,N,N',N'-tetramethyl-ethylenediamine (TEMED) serves as catalyst. A schematic representation of the reaction sequence is shown in Fig. 5-35. In addition to the methacrylic acid group, vinyl silanes can be attached in the first step to the capillary surface by silylation or by treating the chlorinated surface with a Grignard reagent.

The electroosmotic flow is completely suppressed by the polyacrylamide coating. Such capillaries manifest outstanding stability in the pH range of 2.5 - 8. Fig. 5-36 shows a stability test of such capillaries at pH 2.8. Only after 100 separations does a slight deterioration in the resolution become noticeable. However, the capillaries are relatively unstable above pH 8. This becomes noticeable through the gradual reappearance of the electroosmotic flow and fluctuating migration times of the analytes. More recent investigations show that the instability can be caused by hydrolysis of the acrylamide. N-substituted acrylamides are said to yield substantially more pH stable coatings. Vinyl modified capillaries were stable for longer periods of time even at pH 10. On account of their simple preparation or commercial availability there is a wide variety of applications for linear polyacrylamide coatings. These involve primarily the separations of proteins but also include other biomolecules such as DNA fragments. Since linear polyacrylamide coatings can completely suppress the EOF, other classical separation techniques such as isoelectric focusing (IEF) may be adapted to capillary electrophoresis. Capillaries coated with linear polyacrylamide are also used for the preparation of gel-filled capillaries.

Aside from proteins, substances from the area of biologically important samples include primarily nucleotides, oligonucleotides, and DNA fragments. Until now, these types of samples have been separated by MEKC, or preferentially using polyacrylamide gel capillary electrophoresis (CGE). Linear polyacrylamide coatings in combination with macromolecular buffer additives can be used for the separation of DNA restriction fragments over a wide range of sizes. The use of coated capillaries for these applications may prove beneficial in the long run because the problems with the gel-filled capillaries used until now (bubble formation, reproducibility, and gel stability) have not yet been overcome satisfactorily.

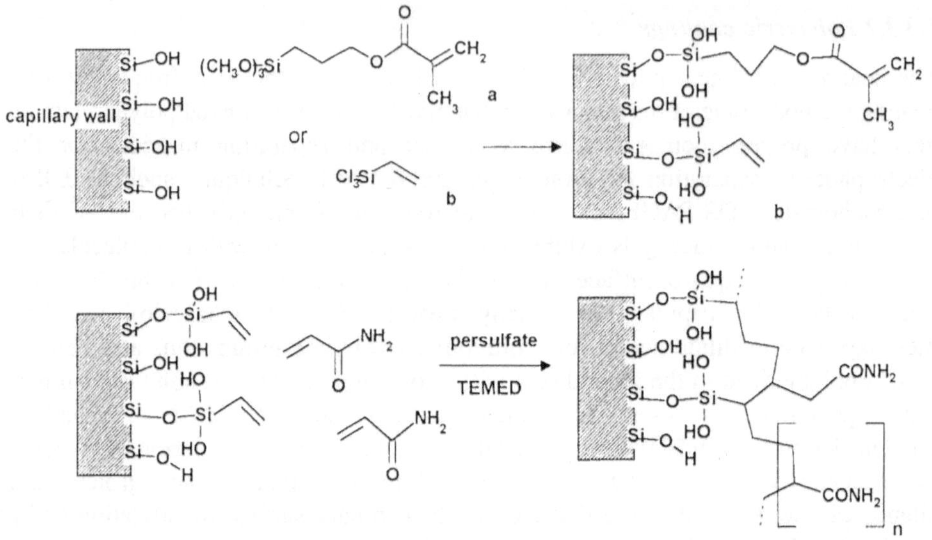

Fig. 5-35 Reaction scheme for the preparation of polyacryamide coatings

1. Attachment of vinyl anchor groups on the quartz surface with γ-methacryloxypropyltrimethoxysilane or trichlorovinylsilane
2. Copolymerization of the vinyl groups with acrylamide by addition of radical initiators

Poly(vinylpyrrolidone) (PVP) coated silica [72] has already been used successfully for the separation of proteins by SEC and HIC in HPLC. The use of PVP as a capillary coating has also been considered for CE [56]. For the preparation, the capillary wall is converted to a vinyl capillary in the first step. Subsequently copolymerization was effected with 1-vinyl-2-pyrrolidone in aqueous solution with TEMED as catalyst and ammonium persulfate as radical initiator. The optimal working range for protein separation for this type of coating lies in a strongly acidic medium where the capillaries are stable for weeks. Fig. 5-37 depicts a separation of 15 standard proteins over a pI range of 4.5 to 11 and molecular weights from 12,000 to 77,000. Efficiencies of up to 700,000 theoretical plates were achieved in a separation requiring less than 25 min. The PVP coating for quartz capillaries has shown itself to be highly efficient and, when used in an acidic pH, entirely comparable to the linear polyacrylamide coating.

Coating with poly(ethyleneimine) yields an ionic product [65]. The *poly-(ethyleneimine)* is attached to the capillary wall through electrostatic interactions as a positively charged polymer. This coating is prepared analogous to the anion-exchange materials based on polymers or silica for HPLC. Prepolymeric polyethyleneimine (PEI) of various molecular weights and concentrations are adsorbed on the carrier surface and subsequently cross-linked.

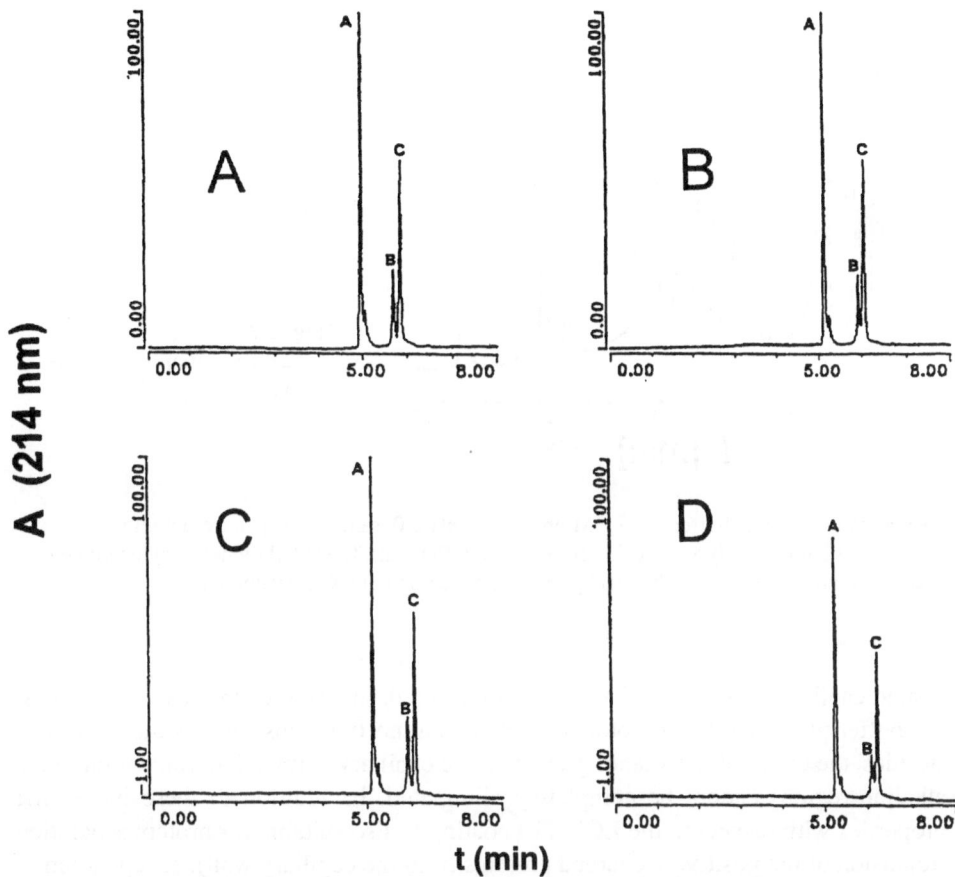

Fig. 5-36 Separation of protein standards in a stability test in a polyacrylamide-coated capillary [59]

Separation conditions: buffer: 30 mM citrate pH 2.8; sample: A lysozyme, B RNA, C chymotripsinogen

The cross-linking agent used is ethylene glycol diglicidyl ether (EDGE). As is known, poly(ethyleneimine) possesses branched polymer chains with primary, secondary, and tertiary amino groups in a 1 : 2 : 1 ratio. These functional groups enable both the electrostatic bonding to the capillary wall and the cross-linking. Due to the high concentration of basic amino groups in the polymer, the layer is positively charged over a wide pH range, which leads to a reversal of the flow direction of electroosmosis. The absolute values of the electroosmotic mobilities achieved with these coatings at low pH are of the same order of magnitude as the flow rates of un-

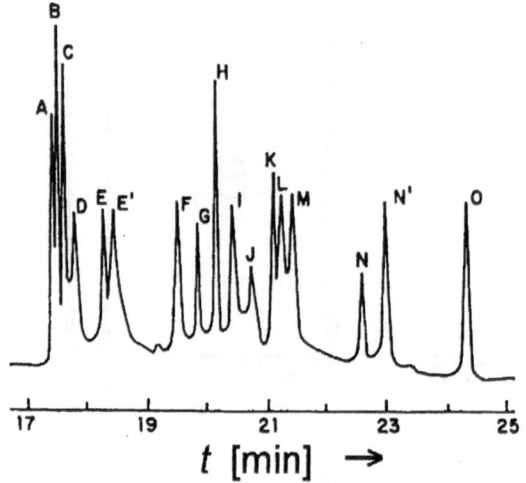

Fig. 5-37
Model separation of proteins
with a PVP-capillary [56]

Separation conditions: buffer: 58.5 mM phosphate, pH 2.0, samples: A β-lactoglobulin B,
B β-lactoglobulin A, C lysozyme, D serum albumin (human), E RNA (bovine), F cytochrome C,
G carboxyanhydrase B, M carboxyanhydrase A, N hemoglobin, O parvalbumin

treated capillaries at basic pH values. As expected, the EOF decreases with increasing buffer pH since the deprotonation of the amino functions follows along with it. Besides, the effect of the silanol groups on the capillary surface (deprotonation in the alkaline region) cannot be completely eliminated. In addition to these interesting properties with respect to the EOF, PEI coating is also suitable for protein separation (repulsion of the positively charged proteins from the capillary wall). A separation of basic proteins at neutral pH is shown in Fig. 5-38.

Polyvinyl-alcohol-coated capillaries [72] exhibit unusual properties and permit highly efficient protein separations. For their preparation, a solution of polyvinyl alcohol is pumped into the capillary and heated at 140°C for several hours. Excess polyvinyl alcohol is removed from the capillary and it is ready for use with an acidic buffer. A stable layer of polyvinyl alcohol has precipitated onto the capillary surface. In the acid and neutral region no measurable EOF is present and the stability of the capillary is very good. The polymeric hydrophilic coating effectively screens the silanol groups of the quartz, and results in separations of extraordinarily high efficiency. An example of a separation is given in 5-39.

Coating with *poly(methylglutamate)* (PMG) [66] can be conducted *in situ* on the capillary surface. The starting material for the preparation of this polypeptide is the monomeric N-carboxy anhydride of α-methylglutamate. This is adsorbed onto the capillary wall and then heated to induce polymerization with the cleavage of carbon dioxide. Due to the very thin coated polymer layer the EOF is reduced only slightly.

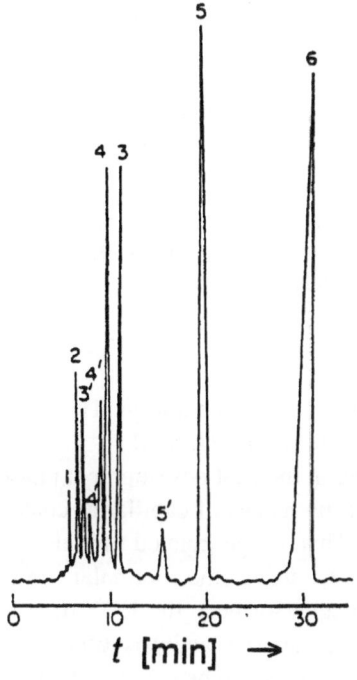

Fig. 5-38
Separation of protein standards in a PEI-coated capillary [65]

Separation conditions: buffer: 20 mM hydroxylamine/HCl, pH 7.0; samples: mesityl oxide (1), myoglobin (whale)(2), BSA (3), chymotripsinogen A (bovine) (4), cytochrome C (equine) (5), lysozyme (6)

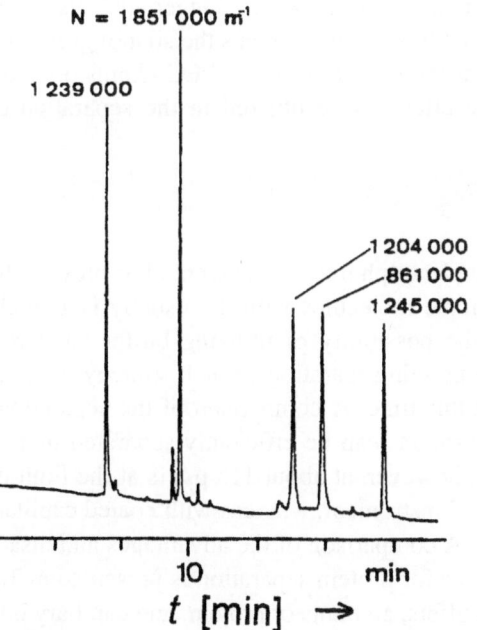

Fig. 5-39
Separation of protein standards in a PVA-coated capillary [72]

Separation conditions: capillary: 50/57 cm, 75 μm i.d., buffer: 50 mM phosphate pH 3.0; field: 430 V cm⁻¹; detection: 214 nm; samples: cytochrome C (1), lysozyme (2), trypsin (3), tripsinogen (4), chymotrypsinogen (5)

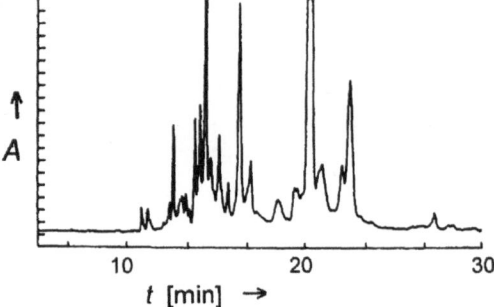

Fig. 5-40
Separation of cellulase-enzyme complexes with a PMG-coated capillary
[66]

Separation conditions: L=34 cm, 75 µm i.d., potential: 8 kV, buffer: 30 mM phosphate pH 7.0

Despite the thin layer, PMG capillaries are suitable for protein separation. The separation of a complex enzyme mixture (of a cellulase) is shown in Fig. 5-40.

Polymer coated GC and SFC capillaries were used in the first development phase of CE. The objective initially was to control the EOF by means of capillaries coated with polymethylsiloxane or polyethylene glycol. This usage related initially to MEKC, and less to the separation of proteins. In using the nonpolar polymethylsiloxane coatings in MEKC (with SDS as micelle former), the EOF rises in comparison to the uncoated capillaries. This leads to shorter analysis times, but also to a lower peak capacity. The increase in electroosmosis can be explained on the basis of adsorption of the surfactant molecules on the hydrophobic surface and therefore with an increase of the negative surface charges. With polyethylene glycol capillaries an opposite effect was introduced, i.e., a lower flow rate, longer analysis times, and higher peak capacities. This means that this coating screens the silanol groups on the capillary wall but because of the hydrophilic nature of the PEG chains no additional adsorption of SDS occurs. These effects were utilized in the separation of purine derivatives and nucleotide bases.

5.3.4 Summary

In the separation of proteins by capillary electrophoresis it is essential to prevent the interaction of the positively charged sample molecules with the usually negatively charged capillary wall. In addition to the possibility of utilizing buffer additives, there are specially coated capillaries for protein separation. Which strategy is better or more sensible cannot be answered at this time. A comparison of the separations of some proteins showed that standard proteins can be efficiently separated in systems using buffer additives. The current, however, at about 115 µA is at the limit of what should be routinely expected of a CE instrument, whereas with coated capillaries more rapid analyses can be achieved. A comparison of the advantages and disadvantages of coated and uncoated capillaries for protein separation is presented in Table 5-9. Care was taken to use the same buffers, amounts of protein, and capillary i.d.

In the last few years, the coating of quartz capillaries has been described by many authors. Until now, however, there are actually very few commercially available coatings (as of January, 1995).

1) Applied Biosystems (Foster City, California) offers a cationic reagent that generates a positively charged coating and thus leads to reversal of the EOF. This coating also reduces the interactions of proteins with the capillary wall at pH values below their isoelectric point.

2) Beckman (Fullerton, CAL) offers capillaries with polar coatings (e.g. polyacrylamide) together with ready to use buffer components for separation of proteins, DNA etc.

3) Bio-Rad (Hercules, California) offers capillaries that have been modified with a hydrophilic coating. This is supposed to reduce the EOF and the adsorption of biomolecules. According to the manufacturer's data, the stability is good even at pH 12, which makes these capillaries suitable for isoelectric focusing.

4) Hewlett Packard (Waldbronn, Germany) offers capillaries with a polyvinylalcohol coating for the separation of biopolymers. These capillaries can also be used advantegeously for the separation of small molecules in CZE.

5) Supelco, Inc. (Bellefonte, Pennsylvania) offers various bonded phases. Besides reducing protein adsorption, these phases are supposed to assure a constant EOF

Table 5-9 Comparison of coated and uncoated capillaries used in CE

Uncoated capillaries	Coated capillaries
1) Cathode-directed electroosmosis	1) Control of electroosmosis (direction, magnitude)
2) Problems with reproducibility of the EOF (hysteresis during pH change)	2) Better reproducibility of the EOF (small hysteresis)
3) strong adsorption of biomolecules (operation with buffer additives is necessary)	3) Passivation of the capillary wall to analytes (repression of wall adsorption)
4) requires careful buffer preparation	4) Problems in reproducing the preparation and in the long-term stability of the coating
5) require small i.d. and low field strength due to buffer additives	5) Low ionic strength buffers can be used (larger inner diameter => high sensitivity)

over the pH range of 3 to 10. Commercial phases include: neutral hydrophilic, weakly hydrophobic C1 phase, hydrophobic C8 phase, strongly hydrophobic C18 phase.

6) Isco, Inc. (Lincoln, Nebraska) offers three coated phases: a covalent bonded C18 phase for proteins, a glycerol coating for proteins and peptides, and a sulfonic acid coating for nucleotides.

7) Various polymeric GC capillaries that are suitable for CE, are also commercially available.

The fact that there are not more vendors of coated capillaries shows that many of the coatings do not meet the requirements of routine analysis. Thus, primarily the wall coatings prepared by conventional techniques are hydrolytically too unstable especially because many separations in CE are carried out in a strongly alkaline medium. Moreover, the coverage of the active sorption sites on the capillary wall is not adequate so that its use for proteins has limited feasibility.

Polymeric coatings are distinctly better in this respect as numerous successful protein separations demonstrate, but even here complete stability to hydrolysis above pH 9 is not assured.

With respect to the separation of biomolecules, it turns out furthermore that none of the wall coatings known thus far exhibits equally good properties for a larger number of different samples (proteins), so that the objective of further research must and will be the development of more stable and universally applicable coatings.

6 MICELLAR ELECTROKINETIC CHROMATOGRAPHY (MEKC)

In CZE the uncharged analytes reach the detector between the anions and the cations without being separated. Micellar electrokinetic chromatography (MEKC), described by Terabe et al. [73] in 1984 enables the separation of uncharged substances through different residence probabilities in the aqueous mobile and a pseudostationary phase.

6.1 Fundamentals of MEKC

By adding detergents to the buffer, micelles are formed after the so-called critical micellar concentration is exceeded. These micelles have a hydrophobic nature on the inside and are charged on the outside which results in an electrophoretic mobility in an electric field. Depending on the nature of the charges, the electrophoretic mobility is directed either toward the cathode or anode. MEKC can be performed in the same apparatus as CZE and requires only the addition of a detergent to the buffer. SDS (sodium dodecyl sulfate) is used most frequently. The resulting micelles carry negative charges and, hence, possess an electrophoretic mobility in the direction of the anode. The analytes distribute themselves between the buffer (transport with the EOF to the cathode) and the inside of the micelle (transport to the anode) so that one can speak of a true chromatographic process. In analogy to CZE, the effective migration velocity of the analytes and that of the micelles is the result of the vectorial sum of the electrophoretic migration and the electroosmotic velocity. Fig. 6-1 [74] represents the separation scheme by micellar electrokinetic chromatography. It involves the most common case in which an anionic detergent is dissolved in a neutral or alkaline buffer.

The electroosmotic flow is directed toward the cathode. When the contribution of the electrophoretic mobility of the micelles is smaller than that of the electroosmotic flow the micelles are transported in the direction of the cathode towards the detector. Very polar molecules that reside only in the aqueous phase migrate at the velocity of the EOF flow and reach the detector at t_o, which, in analogy to LC, is also called the dead time. Strongly hydrophobic analytes stay primarily inside the micelles, migrating at their rate and reaching the detector at t_{MC}. Retarded analytes consequently appear between t_o and t_{MC} at the detector. The actual separation of neutral substances is

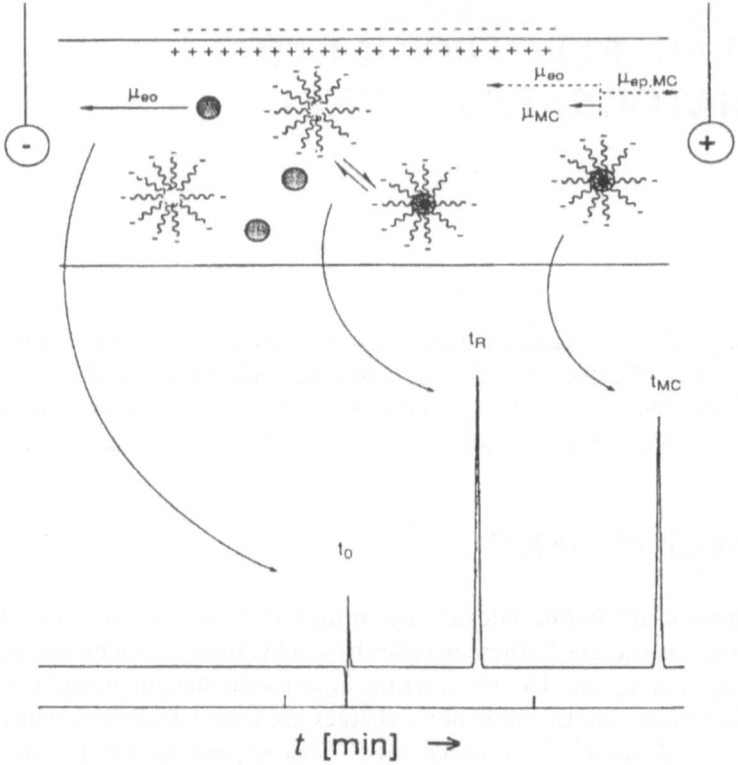

Fig. 6-1 Schematic representation of separation by micellar electrokinetic chromatography [74]

therefore based on their distribution between the buffer solution and the interior of the micelles. Since the separation of the analytes is based on interaction with a pseudostationary phase, one can properly speak of a chromatographic process.

The k' value is defined in analogy to chromatography as the ratio of the residence times of the analyte in the mobile and stationary phases. Taking into consideration that the "stationary" phase moves, the k' value is expressed as

$$k' = \frac{t_R - t_0}{t_0 \cdot \left(1 - \dfrac{t_R}{t_{MC}}\right)} \tag{6.1}$$

In contrast to HPLC, analytes with a k' value of infinity can pass through the detector. In such case the analyte resides exclusively within the micelles. The effect of the distribution of the analyte ions between the pseudostationary phase and the buffer on the separation of a mixture is presented in Fig. 6-2. Both the time and the capacity factors are plotted on separate abscissas.

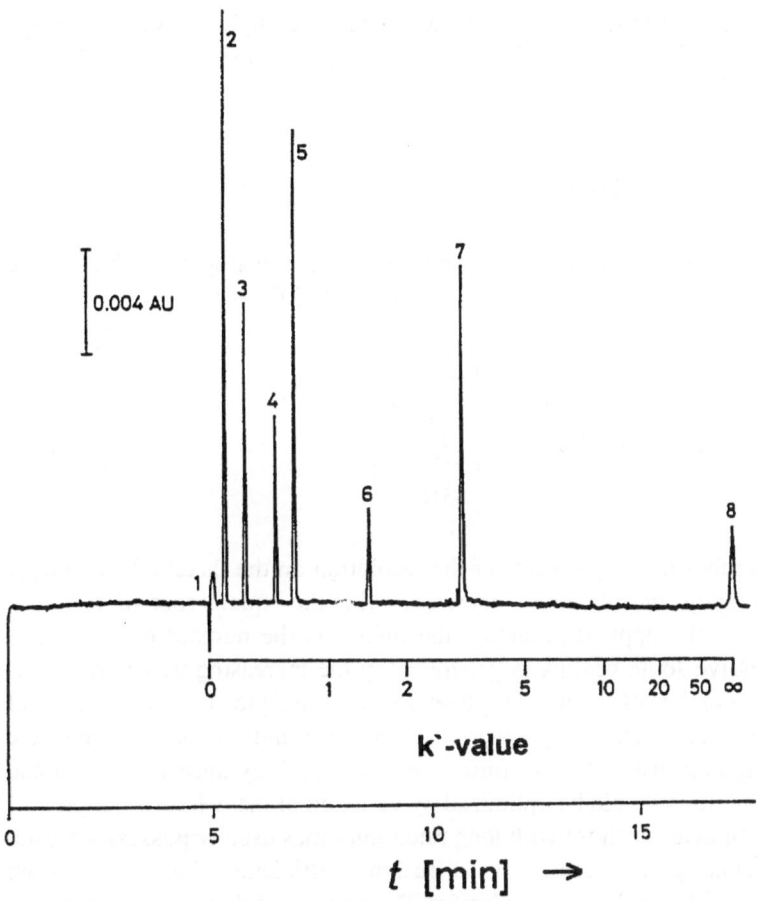

Fig. 6-2 Relationship between migration times and k' values [75]

Separation conditions: capillary: 50/65 cm with 50 μm i.d., buffer: 50 mM phosphate, 100 mM borate, 30 mM SDS, pH 7.0, detection 210 nm, analytes: (1) methanol, (2) resorcinol, (3) phenol, (4) p-nitroaniline, (5) nitrobenzene, (6) toluene, (7) 2-naphthol, (8) Sudan III

For a constant k' the distances between peaks decrease as the k' values increase. To calculate the k' value t_o and t_{MC} must be known. However, there are no ideal and unequivocal markers in MEKC. The marker for t_o must be electrically neutral and completely excluded from the micelles. Suitable inert markers include e.g., acetone, formamide, or methanol, whose interactions with micelles are negligible and which therefore move with the velocity of the EOF. Moreover, these substances can be detected by UV absorption or as a refractive index peak at low wavelengths. Benzyl alcohol is not suitable here owing to its hydrophobicity. The migration rate of the micelles is even more difficult to determine. Markers generally used for t_{MC} are water-insoluble compounds that can be assumed to reside only inside the micelle, e.g., Sudan III or Sudan IV.

6.2 Optimization of Resolution

Resolution in MEKC was defined analogous to that of chromatography: Taking the mobility of the micelles into account, the resolution in MEKC [76] is:

$$R_s = \frac{\sqrt{N}}{4} \cdot \frac{\alpha - 1}{\alpha} \cdot \frac{k'_2}{1 + k'_2} \cdot \frac{1 - \dfrac{t_0}{t_{MC}}}{1 + \dfrac{t_0}{t_{MC}} \cdot k'_1} \tag{6.2}$$

The equation describes the dependence of the resolution on the factors N, k', t_0/t_{MC}. As in chromatography, the resolution increases with the square root of the plate number. The higher the applied potential, the higher is the number of theoretical plates, until excessive Joule heating is generated by the increasing transport of current. The average number of theoretical plates for most analytes lies in the range of 100,000 to 300,000. A substantially lower efficiency may indicate adsorption of the analytes at the capillary wall. Prior to further use, the capillary must be flushed and the separation conditions should be optimized by variation of the pH, for example.

Hydrophobic analytes or those with long retention times usually possess a greater number of theoretical plates because the diffusion coefficients of the micelles are smaller than those of the analytes in the buffer. The number of theoretical plates does not depend appreciably on the capillary length, although with shorter capillaries the injection volume must be correspondingly reduced to limit the contribution of volume overloading to band broadening.

The selectivity is the most important factor because it permits the largest improvement in resolution to be achieved. Selectivity is determined by the partition coefficient of a solute pair between the mobile and stationary phases and therefore depends on the chemical properties of the separation system. The resolution is thus affected by a change in the buffer composition as well as by the selection of another detergent. Two solutes with a selectivity of 1.02 can be readily separated by MEKC. Due to the mobility of the pseudostationary phase, resolution does not increase steadily with the k' value but passes through a maximum. This special characteristic of MEKC is presented in Fig. 6-3. At constant selectivity, the distances between peak maxima decrease at low and high k' values. It can be shown mathematically that the optimal k' value is equal to $(t_{MC}/t_o)^{1/2}$. In real cases optimal k' values of about 2 are obtained, which are similar to those in HPLC at optimal flow velocities. Curve **a** represents the

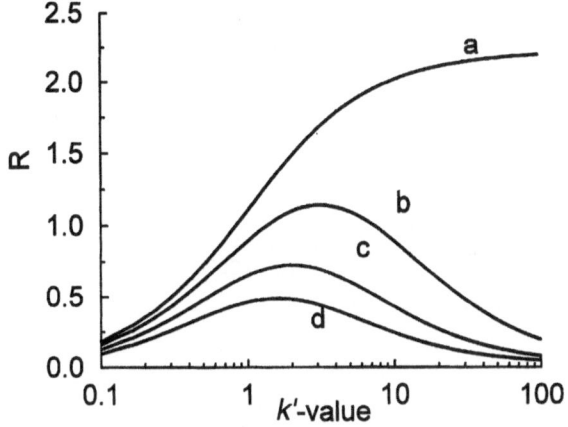

Fig. 6-3
Relationship between resolution and k' value

Calculated resolution of two peaks with N = 200,000, α = 1.02, t_o/t_{MC}: a) 0.0; b) 0.1; c) 0.25; d) 0.4

limiting case when the migration time of the micelle becomes very large. The longer the migration time of the micelle at a given t_o, i.e., the smaller the value for the migration-time window t_o/t_{MC} becomes, the larger is the resolution [77].

Analogous to chromatography, in MEKC a relationship between the k' value and the phase ratio as well as the partition coefficient can be demonstrated:

$$k' = K \cdot V_{MC}/V_{aq,} \tag{6.3}$$

where K is the partition coefficient, V_{MC} the micelle volume and V_{aq} the remaining volume of the buffer. The quotient V_{MC}/V_{aq} represents the phase ratio. In contrast to chromatography, in MEKC the phase ratio can be readily varied via the micelle volume and, hence, the detergent concentration. There is a linear relationship between the k' value and the detergent concentration. Consequently, the k' value can be con-

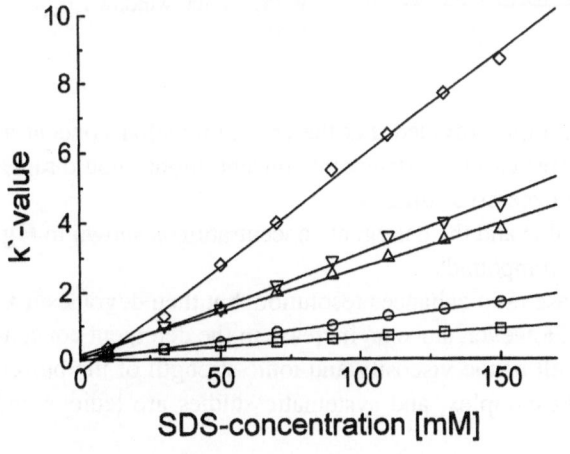

Fig. 6-4
Linear relationship between the k' value and detergent concentration

Separation conditions: capillary: 75 μm x 50/57 cm, buffer: SDS in 50 mM borate (pH 8.5), potential: 20 kV

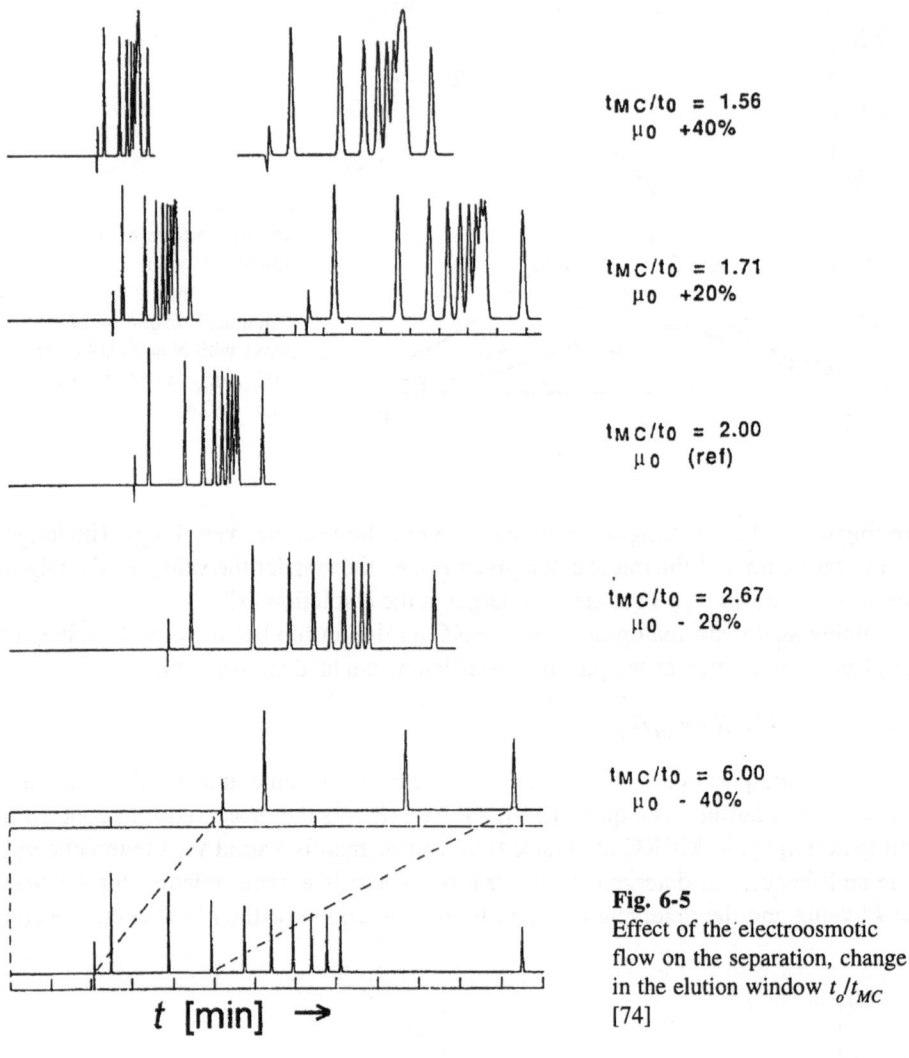

$t_{MC}/t_0 = 1.56$
μ_0 +40%

$t_{MC}/t_0 = 1.71$
μ_0 +20%

$t_{MC}/t_0 = 2.00$
μ_0 (ref)

$t_{MC}/t_0 = 2.67$
μ_0 - 20%

$t_{MC}/t_0 = 6.00$
μ_0 - 40%

t [min] →

Fig. 6-5
Effect of the electroosmotic
flow on the separation, change
in the elution window t_0/t_{MC}
[74]

trolled via the detergent concentration, provided that the critical micellar concentration is known and exceeded. In most cases the detergent concentrations should range between 20 and 200 mM to avoid excessive currents.

The linearity between the k' value and the detergent concentration is shown in Fig. 6-4 for the separation of aromatic compounds.

Initially, an increase in the phase ratio enhances resolution but then degrades it as the phase ratio continues to rise. However, since an increase in the detergent concentration also affects the EOF as well as the viscosity and ionic strength of the buffer, the optimization usually becomes complex, and systematic studies are tedious and time-consuming.

The migration-time window is given by t_o and t_{MC}. The effect of the migration-time ratio t_o/t_{MC} on resolution has just been discussed. The effect of the EOF on the elution-time window is presented systematically in Fig. 6-5. Lowering of the EOF increases the elution-time window and the resolution as well. A disadvantage of this is the hefty extension in the analysis time. In practice, the EOF can be diminished by adding an organic solvent, e.g., methanol or isopropanol (20%). However, this also changes the CMC. Another possibility involves the reduction of the buffer pH. Alternatively, the EOF may be modified by coating the capillary surface or by means of buffer additives. For example, methylcellulose derivatives or ethylene glycol are often added in CE to raise the viscosity of the buffer. However, increased viscosity lowers both the EOF and the electrophoretic mobilities to the same extent. Hence, the resolution cannot be enhanced by raising the viscosity.

The migration-time window can also be extended by increasing the electrophoretic mobility. This method is of little significance in practice, however, because the choice of another detergent would also affect the selectivity. Small changes in the selectivity for small α-values may have a large impact on the resolution, thereby more than negating the expected effect of increasing the electrophoretic mobility of the micelles.

Raising the temperature reduces the migration time because both the partition coefficient and the viscosity decrease. The CMC is also temperature dependent. Since the temperature dependence of the partition coefficients is different for the respective sample components, the selectivity changes. Although a temperature change does not have a large effect on the selectivity, the temperature should be kept constant to ensure analytical reproducibility. Therefore, separation conditions that result in a high current flow should be avoided to prevent heating the buffer and capillary. Efficient cooling of the capillary is therefore beneficial.

6.3 Selection of the Detergent

Ionic detergents are indispensable for MEKC. Many detergents are commercially available, but only a few are suitable for use in MEKC. A detergent for MEKC, must meet the following criteria:
- The solubility in the corresponding buffer must be sufficiently large (>CMC) to permit micelle formation.
- The micellar solution must be homogeneous and UV transparent.
- The micellar solution must possess a low viscosity.

Both anionic and cationic detergents may be employed in MEKC, but the use of the former is far more widespread. As mentioned, SDS (sodium dodecyl sulfate) is utilized most frequently. Homologs of SDS are less suitable. Sodium decyl sulfate at the required concentration (50 mM) possesses an excessively high conductance which results in a high current and, consequently, problems from Joule heating. In contrast, sodium tetradecyl sulfate is too sparingly soluble at room temperature, which restricts its use to higher temperatures. Sulfates and sulfonates are preferred to carboxylates because they possess a constant charge over a wide pH range. Table 6-1 presents some detergents with their critical micellar concentration (CMC) and aggregation numbers. The CMC represents the lowest detergent concentration required to form micelles. The aggregation number is the average number of detergent molecules that assemble to form a micelle. The counter ion of the ionic detergent affects the CMC at a given temperature. For example, sodium dodecyl sulfate is more soluble in water than the potassium salt. In the presence of potassium ions in the buffer an exchange of counter ions may reduce the solubility of the detergent to the extent that the CMC is no longer attained.

Ammonium salts with hydrophobic alkyl chains are primarily used as cationic detergents. The adsorption of detergent molecules at the capillary wall reverses the electroosmotic flow even at concentrations below the CMC, causing the analytes to

Table 6-1 Critical micellar concentrations (CMC) and aggregation numbers of some detergents

Detergent	CMC [mM]	n_{aggr}
Sodium decyl sulfate	33	41
Sodium dodecyl sulfate	8.2	64
Sodium tetradecyl sulfate	2.05	80
Sodium hexadecyl sulfate	0.45	100
Sodium laurylmethyl sulfate	8.7	–
Sodium cholate	13	3
Sodium dehydrocholate	10	–
Sodium taurocholate	10	–
Sodium taurodehydrocholate	6	–
Sodium deoxycholate	6	4
Sodium taurodeoxycholate	9	11
Decyltrimethylammonium bromide	65	–
Dodecyltrimethylammonium bromide	15	50
Tetradecyltrimethylammonium bromide	3.5	75
Hexadecyltrimethylammonium bromide	0.92	61

(The data refer to pure water. In buffer solutions the CMC is lower and the aggregation number is larger)

migrate to the anode. The electric field must be applied with the anode on the detector side. Polymer anions such as polyacrylates and dendrimers with negative charges, which already possess a micellar structure, can be used to advantage. Low concentration suffice as the CMC is attained at low concentrations and is, moreover, independent of temperature [79].

The type of surfactant also exerts great influence on the separation. A detergent molecule consists of a hydrophilic and a hydrophobic portion. Since polar analytes also interact with the surface of micelles, the hydrophilic group (ionic part) has a greater influence on the selectivity of the micelle. Sodium dodecyl sulfate and sodium tetradecyl sulfate exhibit similar selectivities, for example, but this is greatly altered on going to sodium N-lauryl-N-methyltaurate (LMT).

A change in selectivity does not require a complete exchange of detergents but can also be achieved by modification of the micelles. Mixed micelles are formed by adding a second detergent. A micelle consisting of an ionic and a nonionic detergent has a smaller effective charge and is larger. This affects the partition coefficient and gives the mixed micelle a lower mobility. Chiral groups in the detergent allow separation of enantiomers, which is discussed in Chapter 7.

6.4 Separations by MEKC

MEKC extends capillary electrophoretic separation methods to the analysis of compounds that are very poorly soluble in water. One of the first separations based on MEKC, published by Terabe [73], involved benzene derivatives and aromatics (Fig. 6-6).

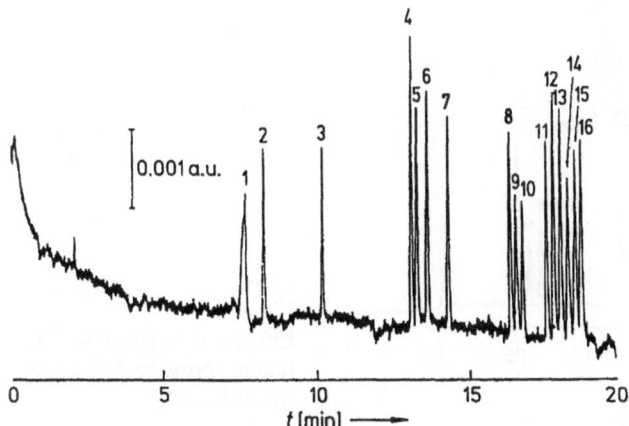

Fig. 6-6
Separation of uncharged aromatic hydrocarbons by MEKC [73]

Separation conditions: 1 mmol SDS in 20 mL borate-phosphate buffer pH 7.0, detection at 270 nm, temperature *ca.* 25°C, E = 300 V/cm, L = 65 cm, 50 μm i.d., analytes: water (1), acetylacetone (2), phenol (3), o-cresol (4), m-cresol (5), p-cresol (6), o-chlorophenol (7), m-chlorophenol (8), p-chlorophenol (9), 2,6-xylenol (10), 2,3-xylenol (11), 2,5-xylenol (12), 3,4-xylenol (13), 3,5-xylenol (14), 2,4-xylenol (15), p-ethylphenol (16)

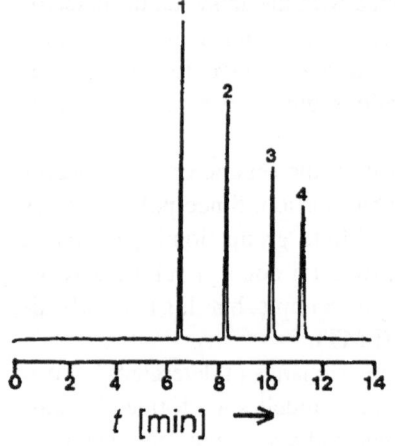

Fig. 6-7
Separation of parabens by MEKC [78]

Separation conditions: capillary: 50/100 cm,
100 μm i.d.; buffer: 100 mM phosphate, 50
mM SDS, pH 7.0; potential: 25 kV; analytes:
methyl, ethyl, propyl, and butyl paraben

The more hydrophobic the analyte, the longer is the residence time in the micelle. Within a homologous series, therefore, the migration time increases, as Fig. 6-7 demonstrates for the separation of a homologous series of parabens (p-hydroxybenzoic acid esters) [78]. These components cannot be separated without the addition of SDS because they migrate with the EOF, i.e., they are eluted at t_o.

The solubility of organic compounds in water can be enhanced by the addition of urea, thereby permitting even steroids to be separated by MEKC. Fig. 6-8 shows the

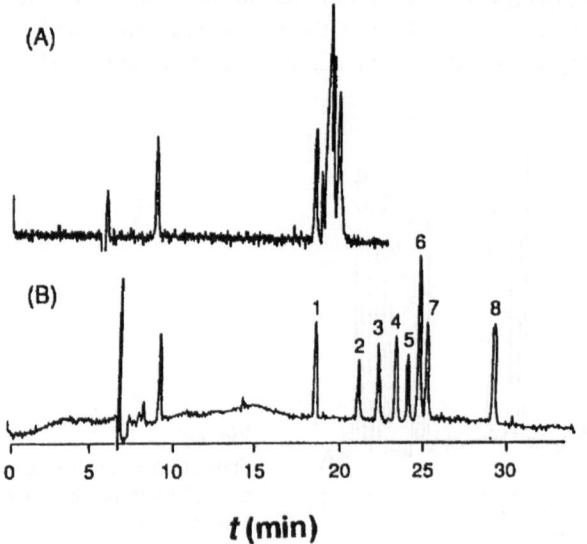

Fig. 6-8
Effect of urea on the separation of cortisone derivatives by MEKC [80]

Separation conditions: capillary: 50/65 cm, 50 μm i.d.; buffer: 20 mM phosphate/borate, pH 9.0, 50 mM SDS; potential: 20 kV; A: without added urea; B: with 6 M urea in the separation buffer; solutes: (1) hydrocortisone, (2) hydrocortisone acetate, (3) betamethasone, (4) cortisone acetate, (5) triamcinalone acetonide, (6) fluocinolone, (7) dexamethasone acetate, (8) fluocinonide

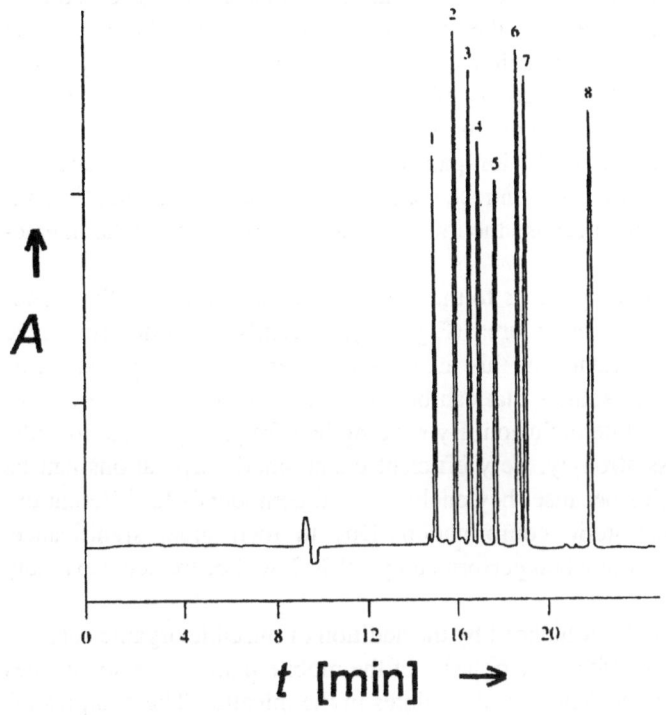

Fig. 6-9
Separation of corticosteroids by MEKC with bile acid as micelle former [80a]

Separation conditions: buffer: 100 mM borate, pH 8.45; 100 mM sodium cholate, potential: 12.5 kV; detection: UV 254 nm, solutes: (1) triamcinolone, (2) hydrocortisone, (3) betamethasone, (4) hydrocortisone acetate, (5) dexamethasone acetate, (6) triamcinalone acetonide, (7) fluocinolone acetonide, (8) fluocinolone

MEKC separation of corticosteroids with and without the addition of urea [80]. A relatively high urea concentration (6 molar) must be used to dissolve the analytes. Corticosteroids can also be separated without the addition of urea by using bile acids as micelle formers [80a] (Fig. 6-9). The similarity of the structure of bile acid to that of the steroids favors their uptake into the micelle. However, the high concentrations permit operation only at relatively low potentials.

The range of application of MEKC is not limited to neutral substances, and charged solutes can be separated as well. The partition of analytes between the aqueous and micellar phases may increase the selectivity and thereby effect by contra-electroosmotic migration the separation of ions with very similar electrophoretic mobilities. In such cases the resolution by MEKC is distinctly better than by CZE.

MEKC may be applied to ionic species where a CZE separation was unsuccessful. The additional partitioning process may make the separation possible. Since micelles are charged on the outside, analytes of the same charge type as the micelles experience stronger repulsion forces than those that are oppositely charged and may additionally form ion pairs. Hence, the partition coefficient is affected by the hydrophobicity and charge of ionic solutes. The addition of ion-pairing reagents to the mobile phase can change the selectivity of ionic compounds considerably. For example, the addition of tetraalkylammonium salts to an SDS solution prolongs the migration time of the anionic analytes by ion-pair formation. The electrostatic repulsion to the micelles decreases as well. In contrast, the migration times of the cationic sample components are shortened because the ion-pair reagent competes for the interaction with the micelles.

The addition of neutral components to the buffer can change the selectivity substantially, as was exemplified with urea (Fig. 6-8). Recently, cyclodextrins have achieved considerably importance in this area. In micellar systems, cyclodextrins function as a third phase for solutes that can be included in them and thereby compete with the micelles in solute inclusion. By residing less frequently in the micelle, the solute is retarded less strongly. Very efficient enantiomeric separations can be achieved with cyclodextrins because they enclose chiral compounds to different extents, depending on their steric conformation. Due to their great significance, enantiomeric separations, even when performed by MEKC, will be treated separately (Chapter 7).

Selectivity is also strongly influenced by the addition of miscible organic solvents to the buffer. This not only affects the polarity of the mobile phase but also changes the EOF and the partition coefficient of the solutes in the micelles. The complex effect of organic modifiers in MEKC is presented schematically in Fig. 6-10. Although

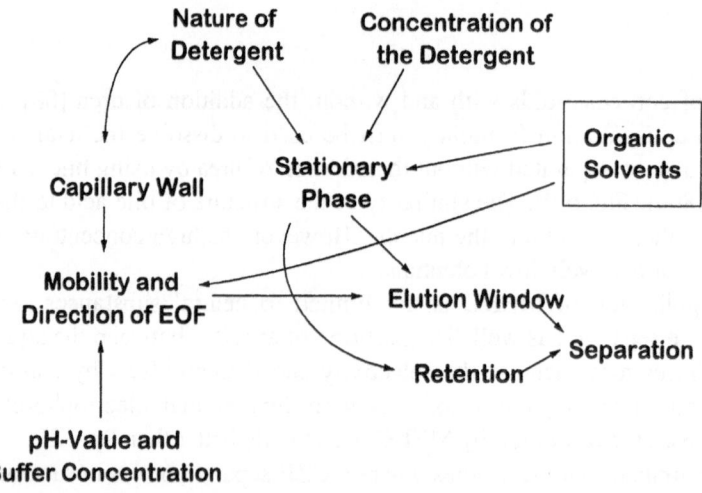

Fig. 6-10 Effect of an organic modifier in MEKC

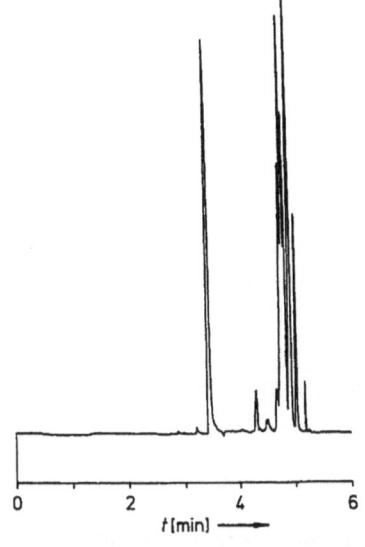

Fig. 6-11
Separation of a mixture of 11 FMOC-amino acids by CZE [6]

Separation conditions: capillary: 50 μm x 50/57 cm, buffer: borate 50 mM pH 9.5; UV detection 200 nm; field 330 Vcm^{-1}

this complexity may give the impression at first glance that MEKC is difficult to use, the high selectivity, together with the large number of theoretical plates achievable, make it a very effective optimization method in CE.

The optimization options in MEKC are illustrated with the separation of the fluorenylmethyloxycarbonyl (FMOC) derivatives of amino acids. Fig. 6-11 shows the separation of 11 FMOC-amino acids by CZE. The relatively large and for all analytes identical contribution from the neutral derivatization reagent leads to very similar electrophoretic mobilities and hence to scarcely any recognizable resolution.

Just by adding SDS to the buffer, substantially better resolution of the FMOC-amino acids is obtained (Fig. 6-12).

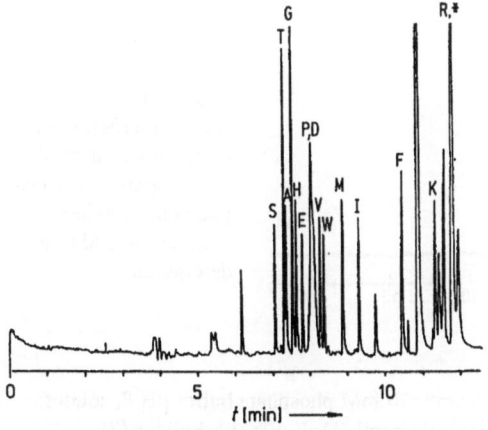

Fig. 6-12
Separation of 16 FMOC-amino acids by MEKC [6]

Separation conditions: buffer: borate 50 mM, 50 mM SDS, pH 9.5, peak identification in the single-letter code for amino acids

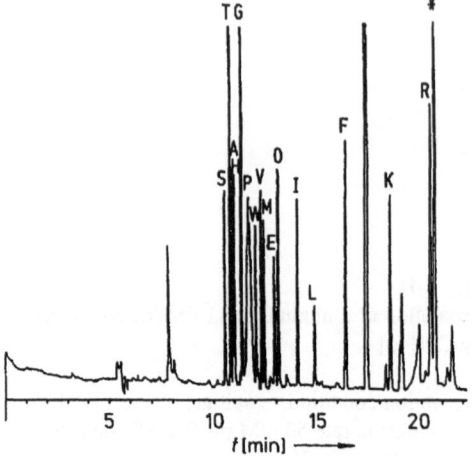

Fig. 6-13
Separation of 16 FMOC-amino acids
by MEKC with methanol as buffer
additive [6]

Separation conditions: buffer: borate
50 mM SDS, pH 9.5, 10 %
methanol (v/v)

The separation can be optimized further by adding an organic component to the buffer. The distribution equilibrium of the analytes between the buffer and the micelle is altered, the EOF is reduced, and the solubility of the analytes in the buffer is improved.

Fig. 6-13 shows the separation of the same sample under identical conditions as in Fig. 6-12, except for the addition of 10 % methanol. These examples demonstrate the options for optimizing the separation efficiency and selectivity in MEKC.

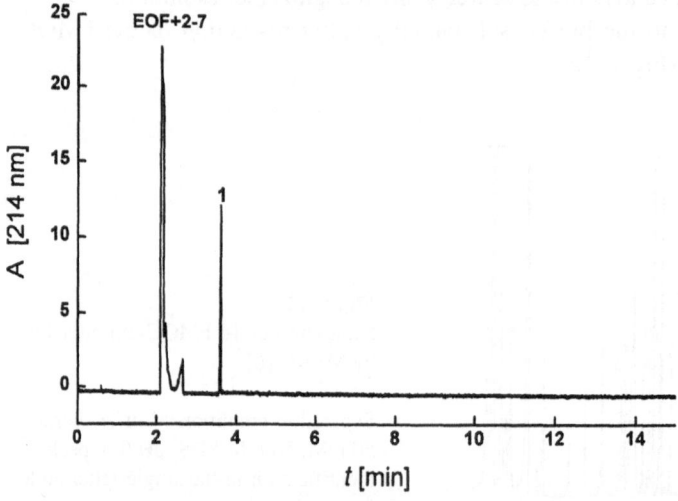

Fig. 6-14
Electropherogram of 7
pesticides (6 neutral
components, 1 carbox-
ylic acid) obtained
without the addition of
detergents

Separation conditions: L = 40/48.5 cm, E = 600 Vcm[-1], 20 mM phosphate buffer pH 8, solutes:
2,4-D (1), Carbendazim (2), Atrazin (3), Diuron (4), Propanil (5), E-605 (6), Folicur (7)

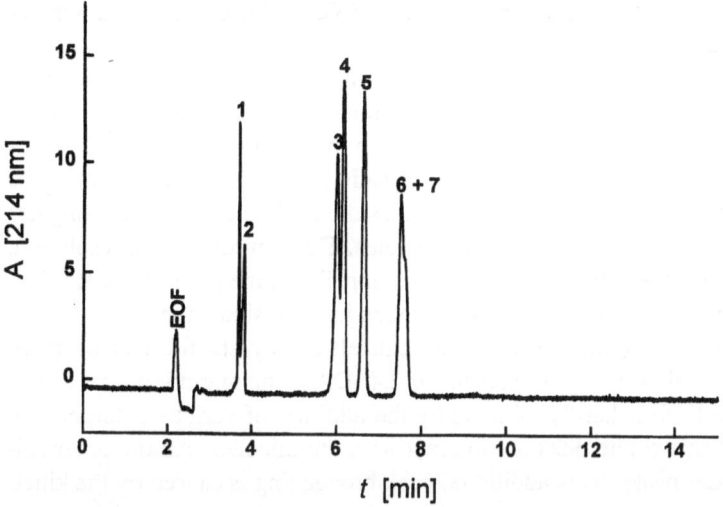

Fig. 6-15 Separation of 7 pesticides in an SDS system

Conditions as in Fig. 6-14 except that 50 mM SDS was added

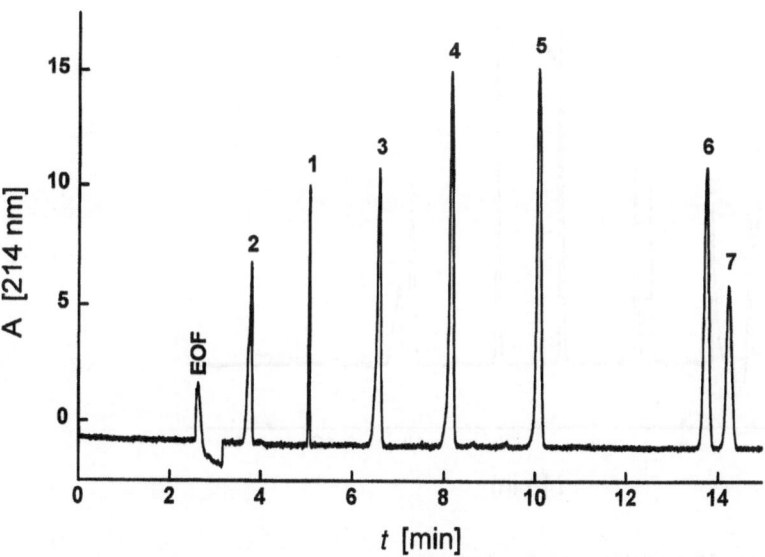

Fig. 6-16 Separation of 7 pesticides in an SDS system with added acetonitrile

Conditions as in Fig. 6-15 except that 10 % acetonitrile was added

The potential for the use of mixed micelles in MEKC is illustrated by the separation of seven pesticides. Fig. 6-14 shows the electropherogram of the sample without the addition of a detergent. The six neutral solutes migrate together with the EOF, and only the carboxylic acid (2,4-D) can be determined. Merely the addition of SDS to the same buffer improves the separation, as shown in Fig. 6-15, although the resolution between all of the components remains unsatisfactory. The addition of 10 % acetonitrile to the otherwise identical buffer reduces the EOF and therefore elongates the migration-time window, as Fig. 6-16 demonstrates. This improves zone resolution. Changes in selectivity caused by the effect of acetonitrile on the partition coefficient and the size of the micelles may also have to be taken into consideration.

The effect of the partitioning process on peak efficiency should also be mentioned. The sharpest peak is always observed for 2,4 -D. As the migration time of this solute relative to the EOF is hardly affected by the addition of various detergents, it can be assumed that this solute does not interact with the micelles. All the other solutes migrate as broader peaks. This additional peak broadening is caused by the kinetics of the partitioning process.

The potential for using mixed micelles is shown in the following figures. In Fig. 6-17 the separation was achieved by the addition of an alkylether sulfate with a C-12 chain. Compared to SDS, better resolution and peak efficiency was obtained but with

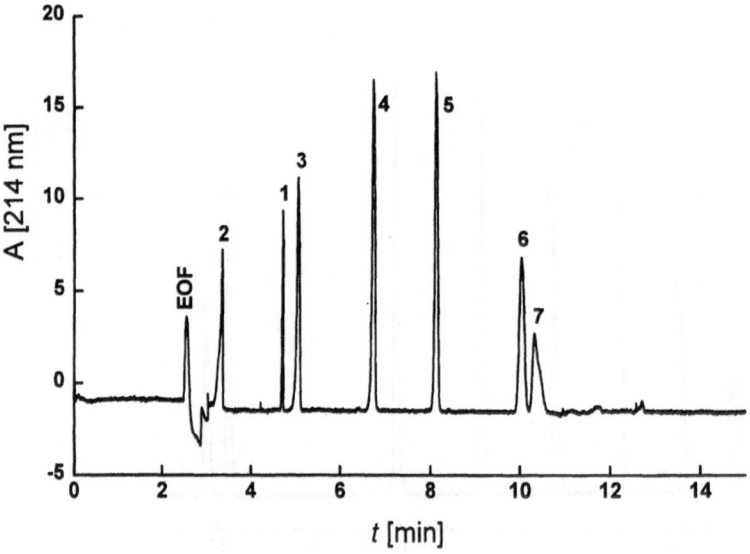

Fig. 6-17 Separation of 7 pesticides with Empicol

Conditions as in Fig. 6-16 except that 50 mM Empicol (ammonium C12-alkylether sulfate) was added in lieu of SDS

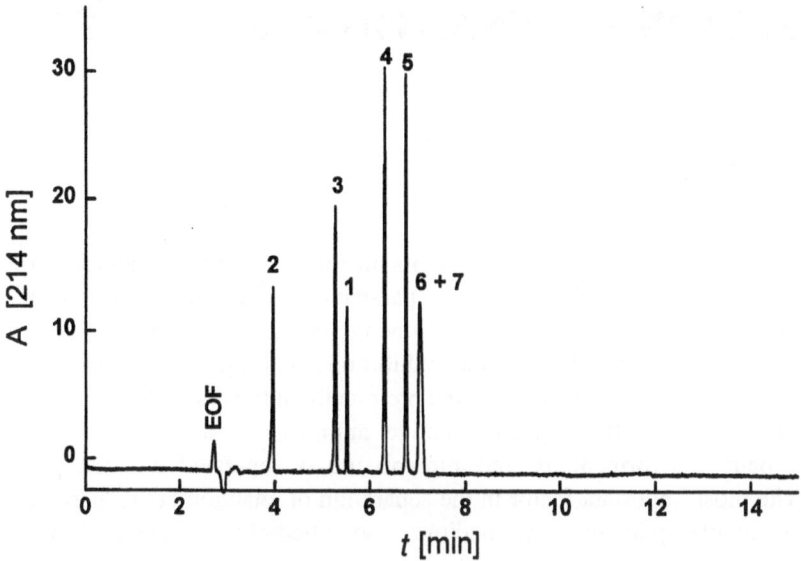

Fig. 6-18 Separation of herbicides via mixed micelles

Conditions as in Fig. 6-16 except that 2 % Renex (C13 fatty alcohol ethoxylate (6EO)) was added

different selectivities. Fig. 6-18 demonstrates the improvements mixed micelles may provide. Here an uncharged detergent was added to the SDS. This alters the size and hydrophobicity of the micelles, as discussed. As can be seen, the resolution and peak efficiency are enhanced.

This example should demonstrate the advantages of MEKC in the separation of nonionic compounds. These applications extend the potential presented by reversed phase chromatography and may serve as an alternative and additional separation system possessing high efficiency but generating different selectivities.

Micellar electrokinetic chromatography also permits the separation of ionic compounds by capillary electrophoresis. It therefore extends the potential applications of electrophoretic separation techniques of neutral and ionic species.

7 SEPARATION OF ENANTIOMERS BY CE

Enantiomers relate to each other as a "picture to its mirror image" and do not differ in their physicochemical properties. They can be separated only in a chiral environment. If the racemic mixture is exposed to an optically active medium consisting of a pure enantiomer, the two analyte enantiomers may exhibit different interactions that leads to their resolution. This may be attributed to the preferential interaction of only one of the enantiomers with the other optically active antipode. If sufficiently stable diastereomeric pairs are formed, the racemic mixture is separated into the pure enantiomers. The most important factor in the separation of enantiomers is therefore the choice of a suitable optically active medium, a so-called chiral selector. Universally applicable chiral selectors are not available and therefore each separation problem must be optimized accordingly.

In capillary electrophoresis the optically active environment is generated by adding a chiral selector to the separation buffer. This simple procedure represents a great advantage for CE, as tedious steps to immobilize the chiral selector unto diverse carriers are unnecessary. As in HPLC, the search for a chiral selector proceeds by the usual trial and error method. A general disadvantage of CE in regard to enantiomer separation stems from its applicability being purely analytical.

Most of the chiral selectors utilized in CE are water soluble and can be conveniently added to the buffer, thereby considerably simplifying the search for the selector with the highest separation efficiency. The chiral selectors used in CE are summarized in Table 7-1 according to their various classes. Most selectors have been used in HPLC and GLC for a long time.

Among selectors, the cyclodextrins and their derivatives afford the broadest and most promising range of applications in CE. The basis for this lies in the molecular structure of many pharmaceutical preparations that form inclusion complexes with cyclodextrins. The optimization of enantiomer separation with cyclodextrins will therefore be discussed in detail. The general structure of the cyclodextrins is shown in Fig. 7-1.

It has been found that other parameters besides the choice of a suitable chiral selector play an important role in enantiomer separation and also require optimization. Since CE separations are based on differences in mobilities, the analytes must be converted to an ionic form by appropriate adjustment of the pH. Neutral and hydrophobic molecules can also be separated via micellar electrokinetic chromatography if they reside for different lengths of time in a hydrophobic phase (SDS in this case). Besides the pH, the most important optimization parameters include the chiral selector con-

Table 7-1 Overview of chiral selectors used in CE

Selector classes	Chiral selectors	Lit.	General remarks
Cyclodextrins	α-, β-, γ-cyclodextrins methylated CD hydroxypropylated CD carboxymethylated CD carboxyethylated CD carboxymethylethylated CD succinylated CD sulfobutyl ether CD aminated and diaminated CD	[81, 82] [81, 83-85] [85, 86] [85, 85a, 87] [85a] [88] [85a] [89] [90, 91]	for solutes capable of forming inclusion complexes; predominant use for aromatic or partially hydro- phobic solutes
Chiral micelle formers (often used with SDS as mixed micelles)	SDS-digitonin SDVal SDAla SDGlu Bile acids	[92] [93-95] [96, 97] [96, 97] [98-100]	in many cases no baseline separation; poor efficiencies if SDVal and SDAla are used alone
Chiral metal complexes	copper-L-histidine complexes copper-aspartame complexes	[101] [102]	separation of dansyl-amino acids; fluorescence detection
Chiral crown ethers	18-crown-6 tetracarboxyl acid	[103, 104]	applicable only to primary basic amines
Pure enantiomers (as ligand exchange reagents)	L-tartaric acid	[105]	separation of chiral metal complexes; few examples of applications
Proteins and glycoproteins	bovine serum albumin, human albumin, orosomucoid, ovomucoid and fungal cellulose, cellulase	[106-108]	detection problems due to UV activity of the chiral selector; only few application examples
Capillary wall coating with a chiral phase	capillary coating with a "Chiral-Dex-Phase"	[69, 70]	only few applications; low phase capacity; stability of the phase

Cyclodextrin	UV transparency	Optical rotation	Solubility (in % w/v at 25° C)	Inner diameter
α-CD	+	+150°	14.5	5.2 A
β-CD	+	+162°	1.8	6.4 A
γ-CD	+	+177°	23.2	8.3 A

Some physical properties of cyclodextrins

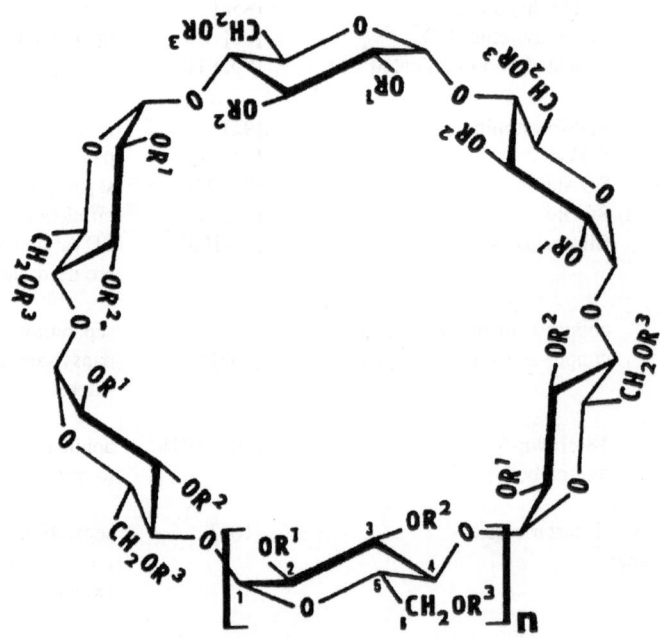

Fig. 7-1 Structure of cyclodextrins

centration in the buffer system, the buffer system itself (type of background electrolyte), other buffer additives such as SDS, organic solvents, urea, etc.

The separation of enantiomers can also be carried out by micellar electrokinetic chromatography. One type of micelle-forming detergents are amino acid derivatives containing a long alkyl chain [93-95]. Their structure is depicted in Fig. 7-2. The other type is bile acid derivatives that have proven their value as chiral selectors in MEKC [98-100]. Their structure is shown in Fig. 7-3.

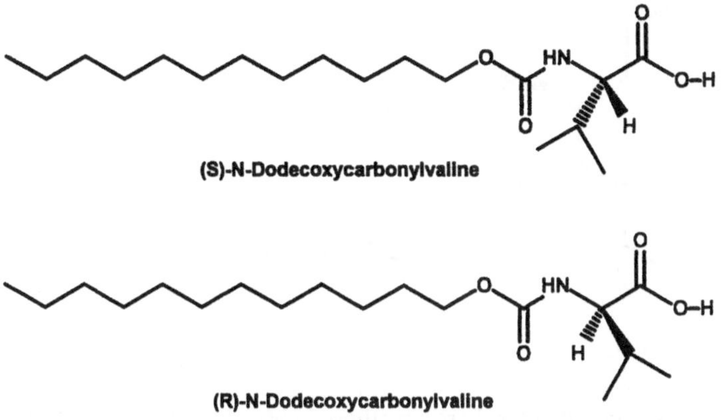

(S)-N-Dodecoxycarbonylvaline

(R)-N-Dodecoxycarbonylvaline

Fig. 7-2 Structure of chiral detergents (long chain amino acid derivatives)

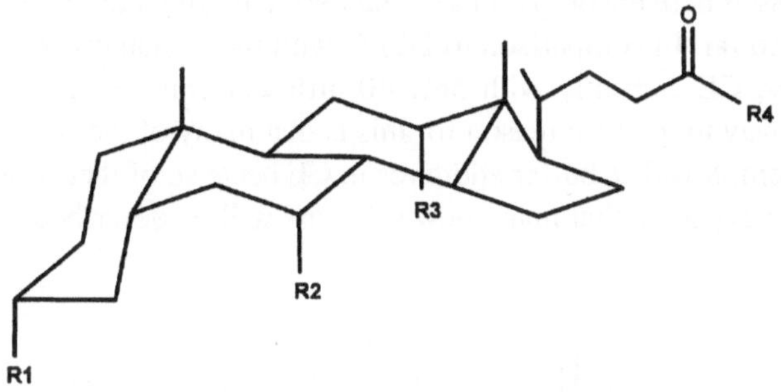

Derivatives of bile acids	R_1	R_2	R_3	R_4	CMC	Aggregation Number
Sodium cholate (SC)	-OH	-OH	-OH	-ONa	13	3
Sodium taurocholate (STC)	-OH	-OH	-OH	-NH-(CH$_2$-)$_2$-SO$_3$Na		
Sodium deoxycholate (SDC)	-OH	-H	-OH	-ONa	6	4
Sodium taurodeoxy cholate (STDC)	-OH	-H	-OH	-NH-(CH$_2$-)$_2$-SO$_3$Na	9	11
Sodium dodecylsulfate (SDS)					8	63

Fig. 7-3 Structure of bile acids

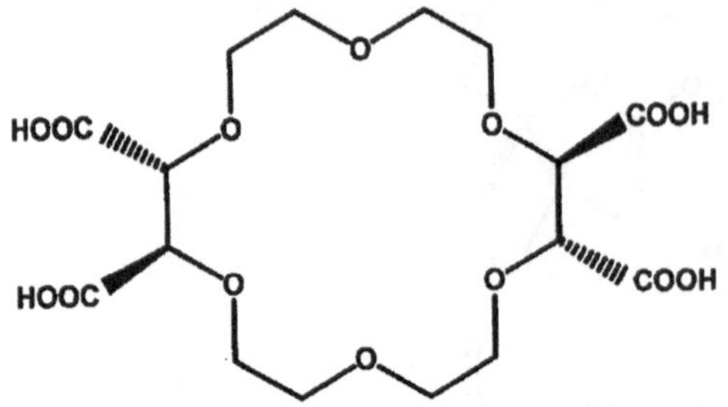

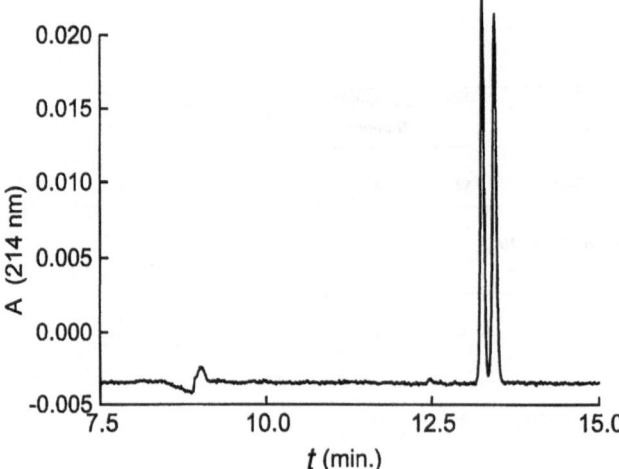

Fig. 7-4 Structure of the crown ethers

Other important selectors include chiral crown ethers, whose structure is presented in Fig 7-4. This class is important because chiral recognition can occur with amines [103, 104]. Many pharmacological substances contain basic functional groups and this may open a broad area of analytical applications for this class of substances.

The selector classes addressed here encompass more than 90 % of all examples of applications described in CE so far. In comparison to HPLC with more than 400 different chiral stationary phases, CE currently with only 40 different chiral selectors may still have a considerable way to go. One reason for this is that many of the selectors used in HPLC cannot be employed as buffer additives in CE because of their UV absorption. However, it can be expected that many new selectors will be described in the future.

Fig. 7-5 Separation of diastereomeric metal complexes without a chiral selector by MEKC

Separation conditions: $L = 60/67$ cm, $E = 370$ Vcm^{-1}, $T = 25$°C, buffer: 16 mM phosphate pH 7.0, 100 mM SDS, 15 % (v/v) methanol, detection at 214 nm

Chiral selectors also assist the separation of diastereomers as well as that of positional isomers. The use of chiral selectors is thus not limited to enantiomeric separations but extends to the entire area of stereoisomeric separations. In many cases the selectivity of other difficult separations may be improved cyclodextrins. In principle, the use of chiral selectors for the separation of diastereomers is not essential. Fig. 7-5 shows the separation of a diastereomeric metal complex by MEKC with SDS as the micelle former, the different hydrophobicities of the diastereomers being the basis of the separation.

7.1 Enantiomeric Separations with Cyclodextrins as Chiral Selectors

Since cyclodextrins (CD) and their derivatives provide the most extensive and promising spectrum of applications as chiral selectors in CE, they will be treated in detail here. Derivatization reactions at the cyclodextrin ring enable the synthesis of a multitude of new types of chiral selectors with a greatly modified effect on the chiral recognition between the selector and the analyte. Table 7-2 summarizes the most important representatives of cyclodextrins used so far. Insofar as they are available, more accurate data on the degree of derivatization, etc. were included. Many of these derivatives are technical products that may vary in the degree of modification and in the position of the functional groups. Therefore, different selectivities may be observed when different batches are used. Products with a defined degree and position of substitution are tedious to synthesize and consequently extremely expensive. For theoretical and mechanistic studies the use of such cyclodextrins may be justified, but for practical separations the purity of the technical products is acceptable. In general, hydrophobic and ionic interactions, as well as steric effects and hydrogen bond formation are responsible for chiral recognition.

The separation of a pair of enantiomers occurs during the electrophoretic migration of the ionic analyte through the "quasistationary phase" (here cyclodextrin). The most important optimization parameters besides the buffer pH are the choice of the chiral selector and its concentration in the buffer system used, the buffer system itself (type of background electrolyte), as well as other buffer additives such as SDS, methanol, etc. Their effect on the separation of enantiomers will be pointed out below.

Since the selectivities are often very low (in general only small differences in mobilities are observed), complete resolution is attained only after relatively long migration distances, even if an appropriate selector was found. Small differences in mobility then lead to a separation of the analytes when the effective migration is maximal. This means that the analyte alone must traverse the longest possible migration distance (from the injection point to the detector). The presence of an EOF adversely affects resolution, especially for analytes with low mobilities, because a strong EOF

Table 7-2 Description of various cyclodextrins

Type of cyclodextrin (CD)	Degree of derivatization (molecules per CD ring)	Molecular-weight (g/mol)	Sum formula
Regular cyclodextrins			
α-CD	-	973	$C_{36}H_{60}O_{30}$
β-CD	-	1135	$C_{42}H_{70}O_{35}$
γ-CD	-	1297	$C_{48}H_{80}O_{40}$
Derivatized cyclodextrins			
Hydroxypropyl-α-CD	3.6	~1180	-
Hydroxypropyl-β-CD	6.3	~1500	-
Hydroxypropyl-γ-CD	4.8	~1580	-
Methyl-β-CD	12.6	~1311	-
Heptakis-(di-O-methyl) β-CD	14	1331	$C_{56}H_{98}O_{35}$
Heptakis-(tri-O-methyl) β-CD	21	1429	$C_{63}H_{112}O_{35}$
Ionizable cyclodextrins			
Carboxymethyl-β-CD	3.6	~1340	-
Carboxyethyl-β-CD	~6	~1570	-
Succinyl-β-CD	3.5	~1490	-
Sulfobutylether-β-CD	4	1730	$C_{58}H_{98}O_{47}S_4Na_4$
Carboxymethylethyl-β-CD	-	-	-
6A-Methylamino-β-CD	1	1165	$C_{43}H_{73}O_{35}N$
6A,6D-Dimethylamino-β-CD	2	1195	$C_{44}H_{76}O_{35}N_2$
2-O-Carboxymethyl-β-cyclodextrin		1193	$C_{43}H_{72}O_{37}$
6[(3-Aminoethyl)-amino]-6-deoxy-β-cyclodextrin	-	-	

does not allow the analytes sufficient time for separation. To attain maximum resolution in the shortest possible time, coated capillaries with a strongly suppressed EOF are recommended. On the other hand, extremely short capillaries (7-20 cm) can also be used with high electric field strengths (up to 1000 V/cm). With a suitable chiral selector, very short analysis times with high resolution can be achieved. The difference between a coated and an uncoated capillary is illustrated in Fig. 7-6 for hexobarbital.

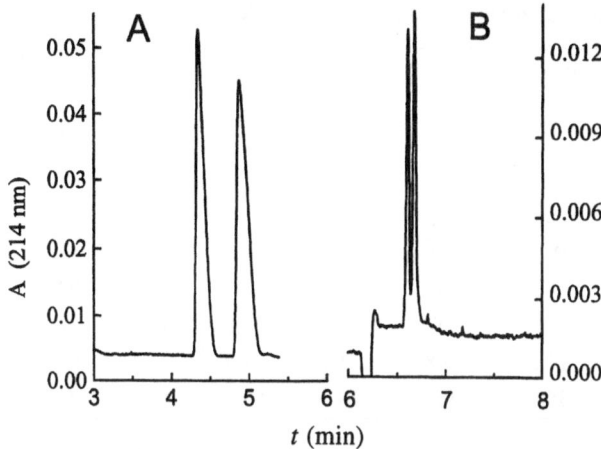

Fig. 7-6 Enantiomer separation of hexobarbital in a coated and uncoated capillary

Separation conditions: 0.1 M TEE pH 8.3, detection at 214 nm, injection: 1s, 2 kV, 1.56 % (w/v)
-cyclodextrin, T = 25°C
A) Coated capillary (4 % T linear polyacrylamide), E = 7/27 cm
B) Uncoated capillary, E = 400 V/cm, L = 50/57 cm

The apparently higher efficiency with the uncoated capillary stems from superimposition of the EOF unto the mobilities of the analytes so that they move very rapidly towards the detector. In addition, the capillary is seven times longer. Consequently, it is not surprising that the plate numbers generated in the uncoated capillary are five times larger (500,000 plates/m) than those in an uncoated one (100,000 plates/m). The enhanced peak resolution in the coated capillary is due to the longer effective migration distance of the analytes (6 cm *vs.* 3 cm).

Table 7-3 summarizes the migration times and effective migration distances for various analytes with different mobilities in relation to that of the EOF. These values were calculated via the following equations:

$$\mu_{Solute} = \mu_{Obs} - \mu_{EOF} = \frac{L_{eff}}{E} \cdot \frac{1}{t_{obs}} - \frac{1}{t_{EOF}} \tag{7.1}$$

$$t_{obs} = \frac{1}{\dfrac{\mu_{Solute} \cdot E}{L_{eff}} - \dfrac{1}{t_{EOF}}} \tag{7.2}$$

$$W_{Solute} = \mu_{Solute} \cdot t_{obs} \cdot E \tag{7.3}$$

Table 7-3 Migration times of anionic analytes at different EOF velocities

	Mobility of solutes as % of EOF							
Mobility of EOF x 10^{-4} $cm^2 V^{-1} s^{-1}$	0	10	20	30	40	50	60	80
8	3.5	3.8	4.3	4.9	5.8	6.9	8.7	17.2
6	4.6	5.1	5.8	6.6	7.7	9.2	11.5	22.8
4	6.9	7.7	8.7	9.9	11.6	13.8	17.3	34.3
2	13.9	15.4	17.3	19.8	23.1	27.8	34.6	68.7
1	27.7	30.8	34.7	39.6	46.2	55.3	69.2	137
Effective migration distance of solute [cm]	0	5.5	12.5	21.4	33.2	49.8	74.7	198

The migration time of the solute, t_{obs}, and that of the EOF, t_{EOF}, were measured for a capillary having an effective length, L_{eff}, under a field strength, E. The migration distance of the solute, W_{Solute}, was calculated for various conditions of the EOF.

As can be seen from the values in this table, a solute exhibiting a mobility of 10 % of the EOF migrates actively only 5.5 cm in a capillary of 50 cm effective length. Most of the transport is effected by the EOF. On the other hand, this means that the same separation should be attainable with a coated capillary having an effective length of 5.5 cm and no EOF as with a 50 cm uncoated capillary. For an analyte approaching 50 % of the mobility of the EOF, its effective migration distance in the capillary is independent of whether a coated or uncoated one is used. The analysis time, however, increases with decreasing mobility of the EOF and increasing mobility of the solute. Long effective migration distances that are necessary for separations with low selectivities always lead to long analysis times. When the mobility of an analyte relative to that of the EOF is known, a rule of thumb can be deduced from Table 7-3 that coated capillaries should be used when the migration time of the analyte is less than twice that of the EOF. Uncoated capillaries give better resolution when the migration time of the analyte is more than twice that of the EOF because of the longer effective migration distance. These considerations for anionic analytes can be correspondingly applied to cationic species as well.

In the search for a suitable chiral selector, the use of coated capillaries is preferred because it permits faster decisions on the utility of a selector, i.e., its selectivity. Of course, in coated capillaries every possible interaction of the analyte with the capillary surface is suppressed, so that higher efficiencies can be obtained. This is particularly noticeable in the separation of strongly basic enantiomers [109]. Because of this fact, enhanced resolution can be achieved with strongly adsorbing solutes as a result of the higher efficiencies.

7.1.1 Neutral cyclodextrins

The choice of a suitable cyclodextrin can only be made by a trial and error process. It is therefore helpful to perform a literature search for similar enantiomers that have already been successfully separated with a certain chiral selector. Not only the different types of CD (α, β, γ) but also their derivatives exert a decisive effect on the selectivity. The solubility of CD in water is greatly improved by derivatization. General rules as to which cyclodextrin makes a separation possible cannot be given as yet, even though many computer model calculations have been performed. Derivatized cyclodextrins exhibit better solubility, especially in the β-cyclodextrin series. Fig. 7-7 demonstrates the selectivity differences of some cyclodextrin derivatives. The hydroxypropyl–β-CD permits separation of both barbiturates and the dansyl amino acids.

The optimization of the cyclodextrin concentration also exerts a decisive influence. According to the type of analyte, enhanced or impaired resolution, or even order inversion may be observed with increasing CD concentration. From this, inferences

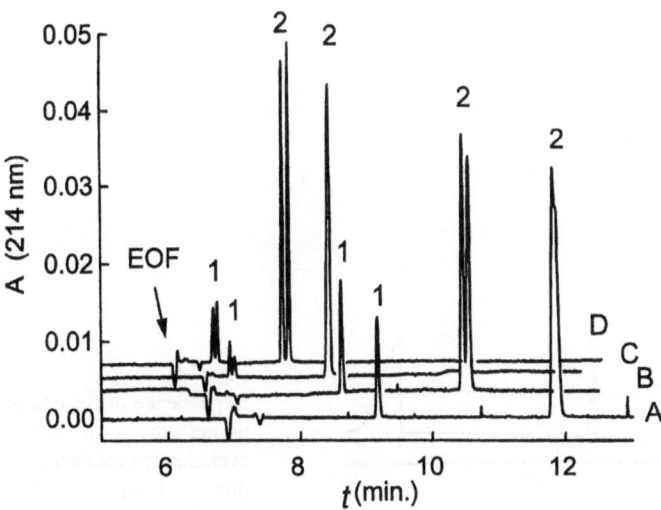

Fig. 7-7 Separation potential of various CD types

Separation conditions: L = 50/57 cm, E = 350 V/cm, buffer: 0.1 M TBE pH 8.3, detection at 214 nm, 1.5 % (w/v) cyclodextrin, sample: 1 d,l-hexobarbital, 2 d,l-dansyl-phenylalanine. A) β-cyclodextrin, B)hydroxypropyl-β-cyclodextrin, C) methyl-β-CD, D) hydroxypropyl-β-CD

can be drawn concerning the different mechanisms for the resolution of enantiomers by means of CD. The following figures (7-8 to 7-10) present the effect of increasing CD concentration on the enantiomer separation. In a buffer system not modified by CD, negatively charged enantiomers possess a relatively high mobility counter to the EOF because they can migrate unhindered through the buffer. The addition of even a small amount of the chiral selector effects a large diminution in the mobilities of the analytes, which results in a separation being observed for the anionic test mixture (Fig. 7-8). This separation at very low CD concentrations may be attributed to the difference in the residence times of the analytes in the CD due to different host-guest stability constants of the D and L forms. The enantiomer with the longest residence time in the CD exhibits the lowest mobility and is detected closer to the EOF.

By raising the concentration of the chiral selector, the mobility of the analytes can be curtailed further, so that they are detected very close to the EOF (Fig. 7-9). In the case of Fig. 7-10 the differences in the stability constants of the enantiomers are too small to permit their separation. Solutes 2 and 3 comigrate and are not separated into their enantiomers, unlike in the case of mandelic acid. Moreover, the buffer viscosity increases at higher CD concentrations, which slows down the EOF and prolongs the analysis times. Even in capillaries with a suppressed EOF, an increase in the cyclodextrin concentration always leads to lower mobilities of the analytes, increased viscosity of the buffer system, and thus also to longer analysis times.

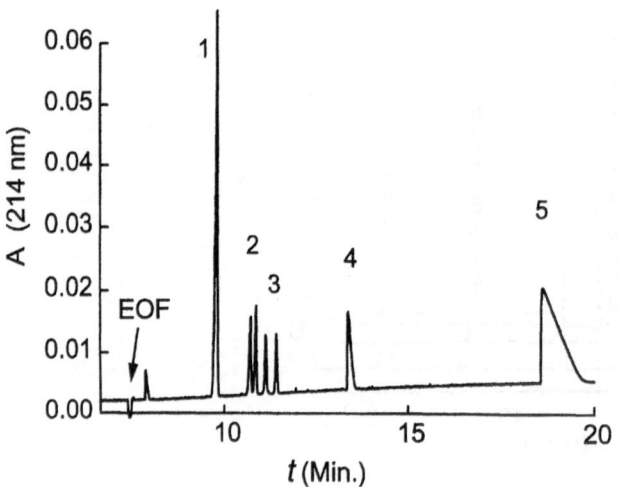

Fig. 7-8
Dependence of the enantiomer separation on the cyclodextrin concentration

Separation conditions: L = 50/57 cm, E = 370 V/cm, 0.1 M Tris/0.1 M boric acid pH 8.3, detection at 214 nm, sample: racemic mixture of hexobarbital (1), a dihydropyridine derivative (2), dansyl-phenylalanine (3), hydrolysis product of hexobarbital (4), and mandelic acid (5), 0.1 % (w/v) hydroxypropyl-β-CD

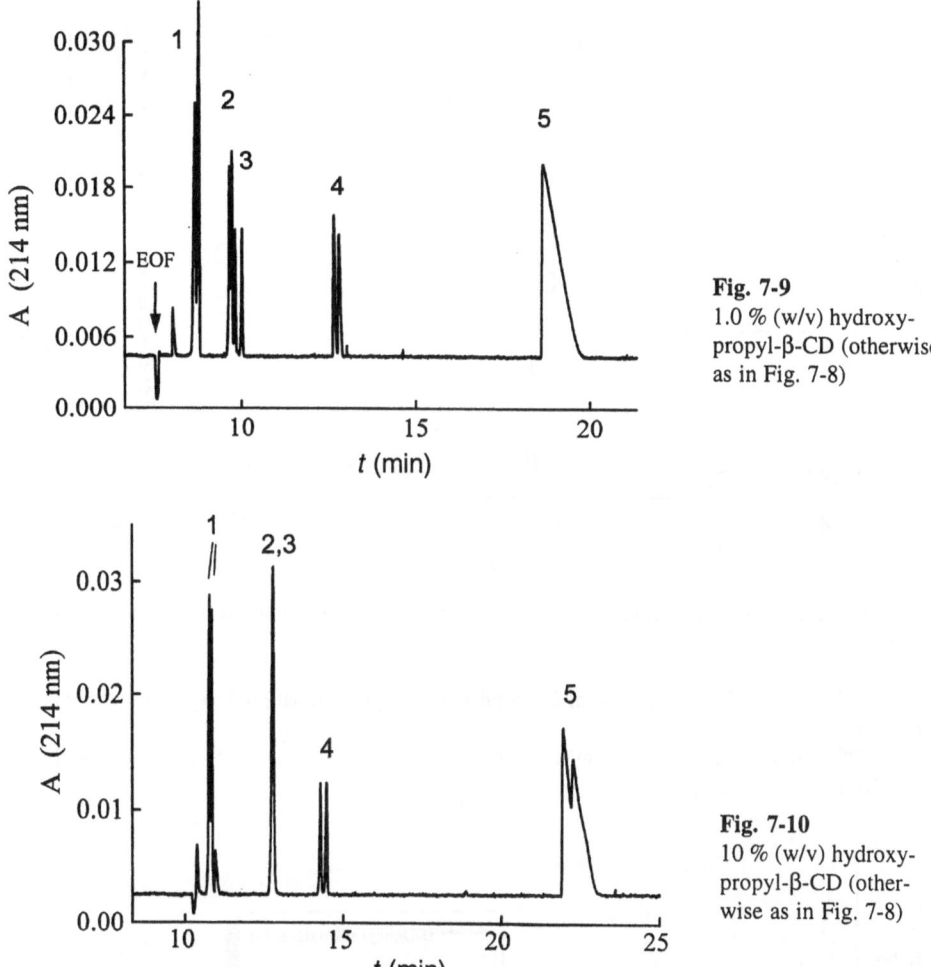

Fig. 7-9
1.0 % (w/v) hydroxy-propyl-β-CD (otherwise as in Fig. 7-8)

Fig. 7-10
10 % (w/v) hydroxy-propyl-β-CD (other-wise as in Fig. 7-8)

Besides the reduction in the resolution due to excessively short residence times in the capillary, other effects are also observed at higher cyclodextrin concentrations. In some cases good resolution (>1.5) of the enantiomeric pair is achieved at very low CD concentrations, but is lost again at higher concentrations of the chiral selector (4-10 % w/v). At higher CD concentrations the residence times of the D and L forms of the analyte increase in the cyclodextrin, but their difference diminishes continuously. This leads to a reduction or even a complete loss in resolution in this system.

An extreme case in the optimization of the CD concentration is presented in Fig. 7-11. Good selectivity is achieved at very low concentrations of the chiral selector. As the concentration is raised, the resolution is lost completely, and returns only at very high CD concentrations. This is also represented in Fig. 7-12 via a plot of the relative migration against the cyclodextrin concentration.

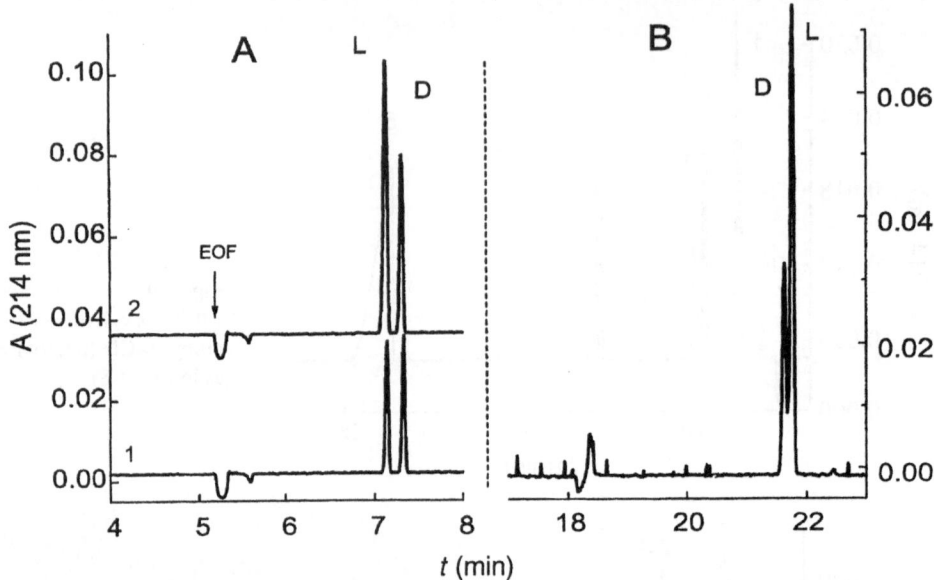

Fig. 7-11 Change in the migration sequence of enantiomers with different cyclodextrin concentrations

A: L = 50/57 cm, E = 350 Vcm^{-1}, 20 mM phosphate buffer pH 7.0, with 0.5 % (w/v) hydroxypropyl-β-CD
B: L = 80/87 cm, E = 300 Vcm^{-1}, 20 mM phosphate buffer pH 7.0, with 15 % (w/v) hydroxypropyl-β-CD, detection at 214 nm

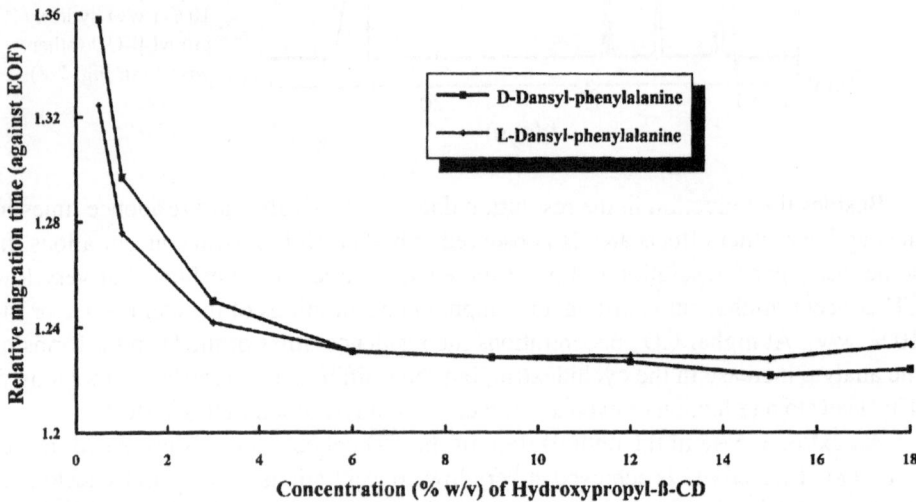

Fig. 7-12 Dependence of the relative migration time of D,L-dansyl-phenylalanine on β-CD concentration (separation conditions as in Fig. 7-11) [85]

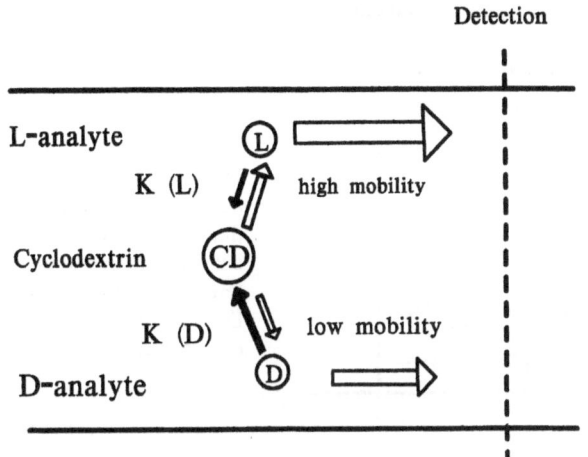

Detection

L-analyte

K (L)   high mobility

Cyclodextrin

K (D)   low mobility

D-analyte

Fig. 7-13
Schematic representation
of the separation of enantiomers due to different
complex formation con-
stants between the D- or
L-analyte to cyclodextrin

It is readily recognized that the relative migration decreases greatly with increasing CD concentration and reaches a minimum at about 6 % (w/v). In this range the
resolution depends only on the difference in the residence times of the D and L forms
of the cyclodextrin, i.e., the differences in the stability constants of the D and L
analytes are responsible here for the separation of the enantiomers. This is represented schematically in Fig. 7-13. When the molecule is in the CD complex, it migrates more slowly as an anion against the EOF. Since all solutes are transported by
the EOF to the detector, the peaks of these anions appear first in the
electropherogram. If the tendency toward complexation is lower, the anions remain
longer in the buffer, migrate faster against the EOF, and are detected later.

From about 6 % CD the mobility of the analyte scarcely changes. Noteworthy,
however, is the reappearance of the resolution at CD concentrations greater than 12
% (w/v). This can only be explained by the formation of diastereomeric complexes
between cyclodextrin and the analyte, and their migration as such. Diastereomers by
nature manifest different physical properties and are therefore separable. As a consequence of the different separation mechanisms at the beginning and at the end of the
curve, the migration sequence of the enantiomers necessarily reverses. This is represented schematically in Fig. 7-14. A higher stability of the complex means a deeper
penetration of the complexes and thus a greater charge density. For this reason, the L-
complex migrates substantially more slowly than the D-complex, which results in
them being resolved again.

The optimization of the pH also represents an important factor. Not only is the
EOF influenced by means of the pH but the analytes are also converted to a definite
ionic form. This gives rise to distinct electrophoretic mobilities that then lead to the
separation of the analytes. If uncharged cyclodextrins are used, the enantiomeric pair

Detection

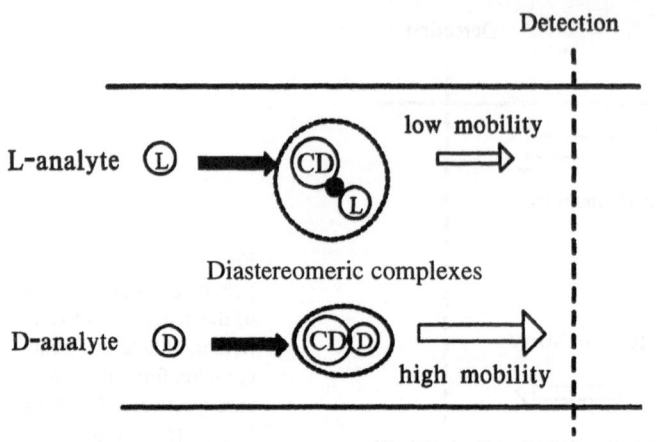

L-analyte

low mobility

Diastereomeric complexes

D-analyte

high mobility

Fig. 7-14
Schematic representation of the separation of enantiomers due to different mobilities of the D-cyclodextrin or L-cyclodextrin complexes

must possess its own mobility at a definite pH in order to be able to migrate through the pseudostationary phase, here the cyclodextrin.

Fig. 7-15 shows the pH dependence of a separation, using dihydropyridine-carboxylic acid as an example. At low pH the analyte possesses scarcely any mobility of its own and is moved past the detector close to the EOF without being separated. The EOF is also very small at low pH. At moderate pH values the analyte itself has a sufficiently large mobility so that the difference between the D and L forms improves selectivity. In this case the analyte has a long enough path length to traverse in the capillary. The loss in resolution at higher pH values may be attributed to the very rapid EOF, leaving the analyte insufficient time for separation through contra-

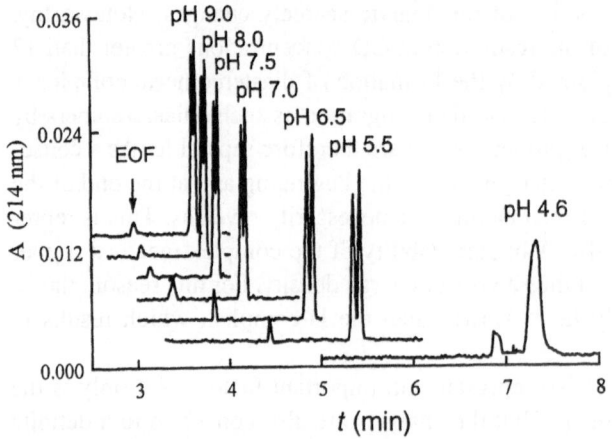

Fig. 7-15
Effect of pH on the separation of a dihydropyridine derivatives

Separation conditions: L = 50/57 cm, E = 440 Vcm⁻¹, buffer: 20 mM phosphate, 0.4 % (w/v) hydroxypropyl-β-CD, different pH values, detection at 214 nm

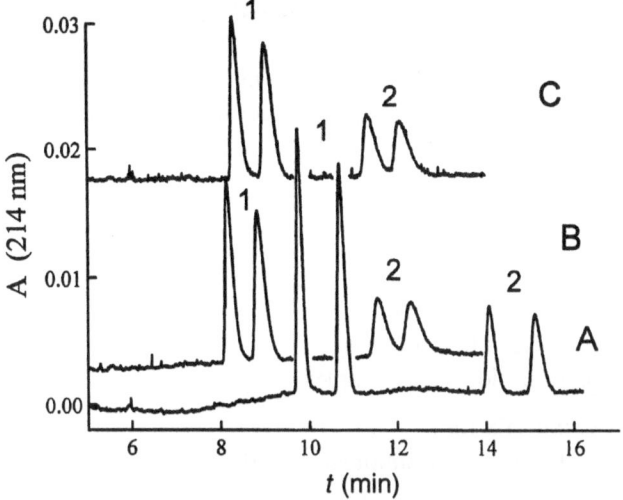

Fig. 7-16
Effect of the background electrolyte on peak shape and resolution [85]

Separation conditions: L = 20/27 cm, polyacrylamide coating, E = 370 Vcm^{-1}, (outlet grounded), 0.3 % (w/v) hydroxypropyl-β-cyclodextrin, detection at 214 nm, sample: dansyl-phenylalanine (1), dihydropyrideine derivative (2),
A) 10 mM citric acid/Tris pH 6.0
B) 25 mM MES/Tris pH 6.0
C) 20 mM phosphate buffer pH 6.0

migration in the capillary. The rapid superimposed EOF thus leads to an excessively short dwell time of the analyte in the capillary. At very high pH values (>12) the cyclodextrins themselves can be deprotonated, which also alters the selectivity. If the cyclodextrin has the same charge as the analyte, chiral recognition is lost on account of electrostatic repulsion.

The optimization of the background electrolyte must lead to an enhancement of the plate number. As was shown in Section 3.4.3 on electrodispersion, the mobility of the background electrolyte should correspond closely to that of the solute. The influence of the type of buffer is demonstrated in Fig. 7-16 for three different buffers. In all three cases the injection and all other separation conditions, except for the buffer ions, were kept constant. It is clearly evident that better resolution was achieved with the citric acid buffer as a consequence of higher efficiency. A higher detection sensitivity for the system was also produced. This example clearly demonstrates that adequate resolution of the analytes may be achieved even for small differences in mobility by improving the efficiency.

In addition to the parameters already discussed, buffer additives also exert a great influence on the separation effectiveness of a chiral selector. However, it cannot be predicted whether the addition of other components (e.g., organic solvents, complexing agents, detergents, etc.) to the buffer will improve the separation or perhaps even completely destroy it.

Some publications have therefore attempted to formulate a model for this situation [110, 111]. The principal factors for the difference in mobilities are presented in the following equation.

$$\Delta\mu = \frac{[C](\mu_1 - \mu_2)(K_B - K_A)}{1 + [C](K_A + K_B) + K_A K_B [C]^2} \tag{7.4}$$

$\Delta\mu$ = difference in mobilities of the enantiomers
μ_1 = mobility of enantiomer 1 or 2 in free solution
μ_2 = mobility of the enantiomer-cyclodextrin complex

$[C]$ = concentration of the chiral selector
K_A, K_B = equilibrium constants between enantiomer A or B and cyclodextrin

The mobility differences and hence the selectivity are predominantly dependent on the cyclodextrin concentration, the equilibrium constants between the analyte and the chiral selector, and the mobility differences between the complexed and uncomplexed form of the analyte. It follows from the above relationship that at constant cyclodextrin concentration the addition of an organic component to the buffer may drive the equilibrium constant to an enhancement as well as to an impairment of the resolution. In which direction it is driven depends predominantly on the cyclodextrin concentration.

How involved the overall processes are becomes clear from Fig. 7-17 via the effect of the additional urea, methanol, and SDS on the migration time, peak efficiency, and resolution. The fastest migration time is obtained in a buffer solution of saturated β-CD (1.56 % w/v) (A). The cyclodextrin concentration can be increased, however, by the addition of 7 molar urea (D). In this case the analysis time rises drastically owing to the greater viscosity of the buffer and the lower mobility of the analyte caused by the high concentration of the chiral selector. However, no improvement is observed in the resolution. The addition of methanol (E) has a positive effect in this case. The migration time increases somewhat, but an enhanced resolution is achieved. By using the buffer from (D) together with 0.1 M SDS the migration times are greatly shortened (B). This stems from the fact that under the existing conditions the SDS and the analyte migrate in the same direction (both are anionic), which gives rise to a synergistic effect. The resolution is greatly improved in comparison to (D) and the analysis time is shortened. Here, too, the addition of methanol to this buffer system prolongs the analysis time, but no improvement in the resolution is observed (C). In the examples discussed here, improved resolution may be ascribed predominantly to higher efficiencies in a certain separation medium, as the migration differences of the enantiomers hardly change.

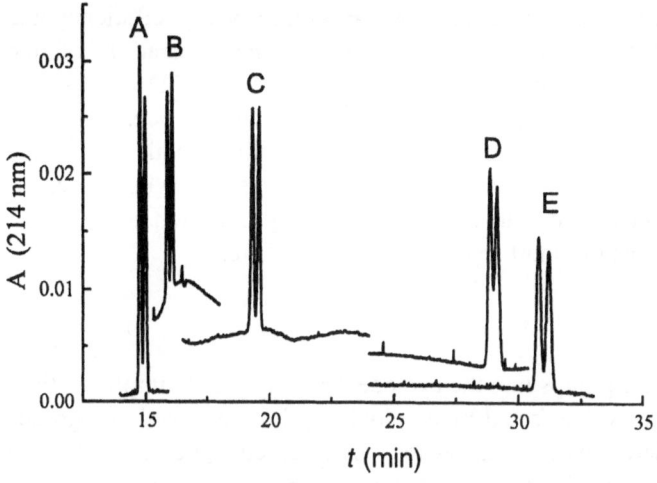

Fig. 7-17
Effect of buffer additives
on the separation of
enantiomers

Separation conditions: L = 40/47 cm, E = 232 Vcm^{-1} (anode at the detection side), coated capillary (4 % T linear polyacrylamide), detection at 214 nm, buffer: 0.1 M TBE pH 8.3 with:
A) 1.56 % (w/v) β-cyclodextrin
B) 12 % (w/v) β-cyclodextrin, 7 M urea, 0.1 % SDS
C) 12 % (w/v) β-cyclodextrin, 7 M urea, 10 % (v/v) methanol, 0.1 M SDS
D) 12 % (w/v) β-cyclodextrin, 7 M urea
E) 12 % (w/v) β-cyclodextrin, 7 M urea, 10 % methanol

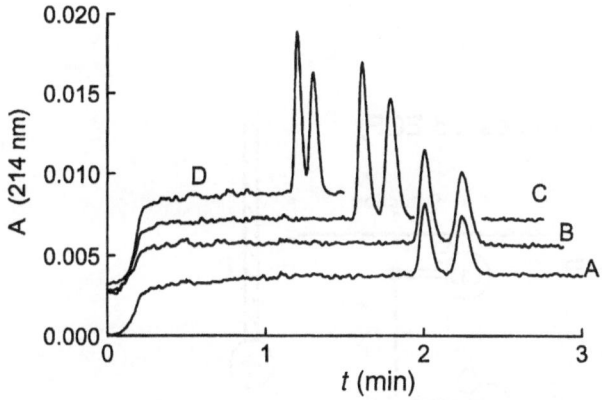

Fig. 7-18
Optimized separation of a
dihydropyridinecarboxylic
acid derivative in CE

Separation conditions: L = 7/27 cm, 75 μm i.d. coated capillary (4%T linear polyacrylamide), T = 25°C, 20 mM phosphate buffer pH 6.0 with 0.5 % (w/v) hydroxypropyl-γ-cyclodextrin, injection and separation potentials:
A) 2s, 3 kV; E = 555 Vcm^{-1}
B) 4s, 4 kV; E = 555 Vcm^{-1}
C) 4s, 4 kV; E = 680 Vcm^{-1}
D) 4s, 4 kV; E = 740 Vcm^{-1}

The potentiality of enantiomeric separation by CE with respect to efficiency and speed of analysis is demonstrated in Fig. 7-18 for the separation of the dihydropyridinedicarboxylic acid from Fig. 7-16. Hydroxypropyl-γ-CD was used as selector in conjunction with an acrylamide-coated capillary. Complete separation was achieved in less than 2 min using a capillary of 7 cm effective length and a field strength of 680 V/cm. With the instrument employed, it was possible to use the capillary portion between the detector and the end of the capillary (injection at the end of the capillary), which produced the short separation distance of 7 cm.

7.1.2 Ionic cyclodextrins

The introduction of ionizable groups into the cyclodextrin ring greatly improves their potential for application. This is effected by employing primarily functional groups such as amino, carboxyl derivatives, or sulfonates which, depending on the pH of the buffer system used, may exist in a protonated or deprotonated, i.e., ionized or nonionized, state. It follows that these chiral selectors have their own mobility in their ionized form. The formation of inclusion complexes with these charged cyclodextrins affords many new possibilities similar to those described for micellar electrokinetic chromatography.

As is shown schematically in Fig. 7-19, a charged cyclodextrin is capable of moving a neutral analyte to the detector. Since such a cyclodextrin has its own mobility, a neutral analyte can acquire a mobility relative to the EOF after forming a host-guest complex. Fig. 7-20 shows the separation of uncharged analytes with a charged cyclodextrin anion.

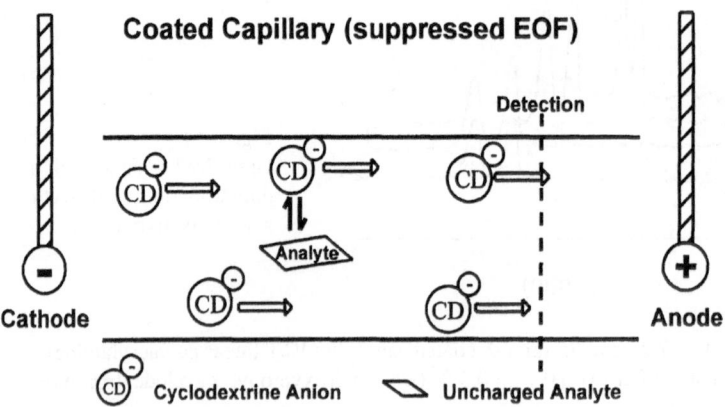

Fig. 7-19 Schematic representation of the transport mechanism of neutral analytes through charged cyclodextrin

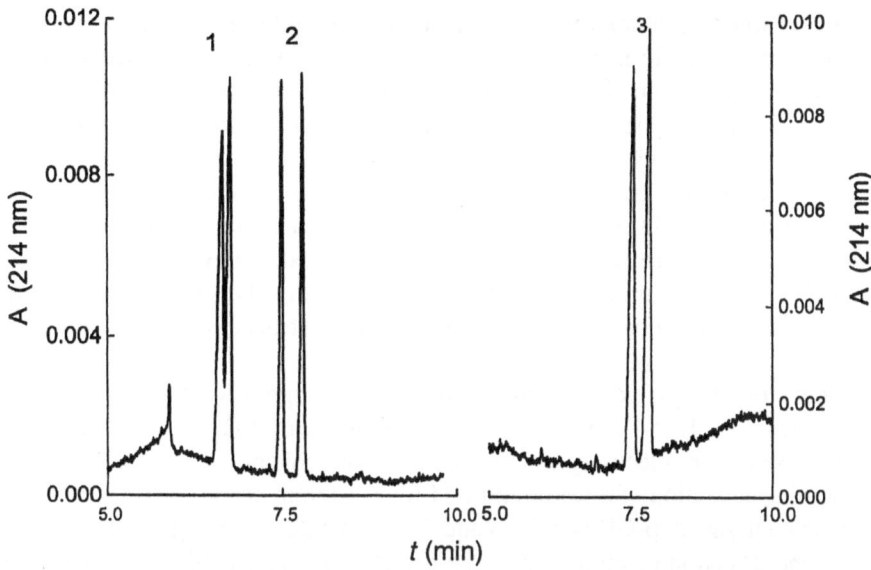

Fig. 7-20 Separation of uncharged enantiomers with charged cyclodextrins

Separation conditions: L = 20/27 cm, 75 μm i.d. coated capillary (polyacrylamide), T = 20°C, E = 400 V/cm, 0.66 % (w/v) carboxyethyl–γ-CD, 20 mM phosphate buffer pH 5.5 (adjusted with Tris), analytes: racemate of hexobarbital (1), oxazolidinon (2), binaphthol (3)

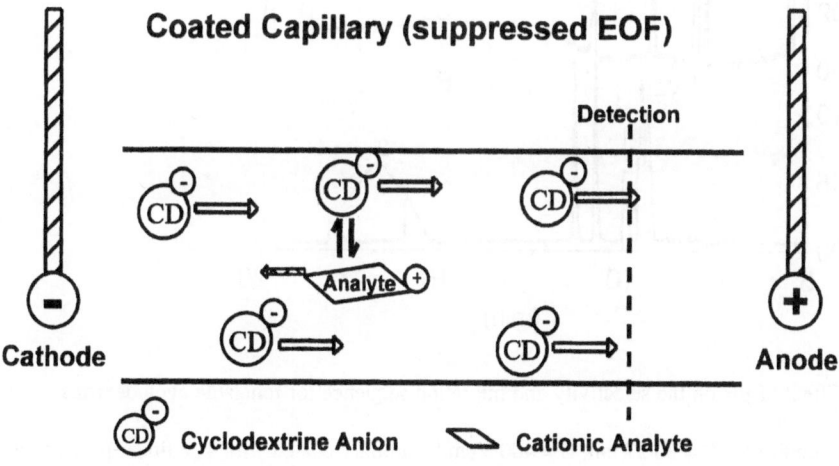

Fig. 7-21 Schematic representation of the direction of migration for cationic enantiomers using multiply charged cyclodextrin anions

Multiple charged cyclodextrin anions create the possibility of an additional ion-pair mechanism that could lead to an altered selectivity and migration sequence of various cationic enantiomers. Under these conditions, the direction of migration of the analyte depends to a large extent on the magnitude of the interaction with the CD. This is shown schematically in Fig. 7-21. Analytes having a high affinity for the mul-tiply-charged CD anion form anionic complexes if their positive charge is overcom-pensated by that of the CD, and this causes them to migrate with the latter to the an-ode. Thus, the cyclodextrin containing "positively charged" analyte migrates to the anode. Analytes with less pronounced affinities for the charged cyclodextrins migrate in accordance with their charge to the corresponding electrode, i.e., cations towards the cathode.

Depending on the pH, cyclodextrins with carboxyl groups can be used in a protonated, neutral (pH < 4), or deprotonated, anionic form (pH > 5). This means a strong influence on the selectivity and on the migration sequence of the analytes. Fig. 7-22 shows the separation of a test mixture of four different enantiomeric molecules with carboxymethylated β-CD at pH 2.5 and 5.5. At pH 2.5 the charge of the analyte determines the direction of migration to the detector, as the cyclodextrin carries no charge and behaves as an uncharged molecule. In contrast, at pH 5.5 the cyclodextrin anion determines the migration direction of the analytes.

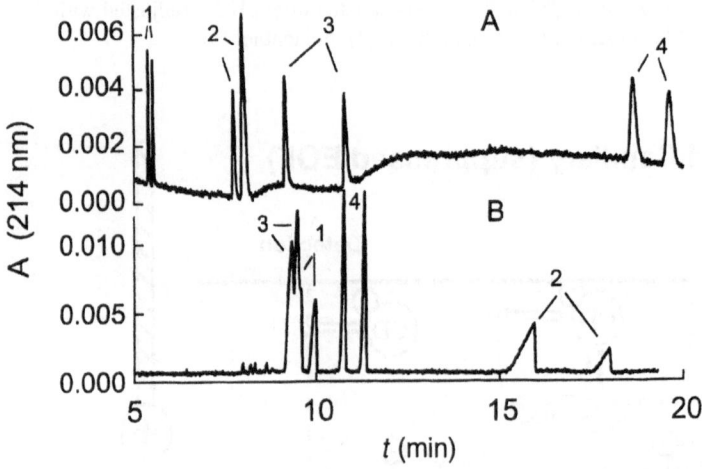

Fig. 7-22 Effect of pH on the selectivity and migration sequence for ionizable cyclodextrins

Separation conditions: L = 30/37 cm, E = 400 Vcm^{-1}, capillary coated with 4 % linear polyacryla-mide, T = 20°C, sample: racemic mixture of doxylamine (1), ephedrine spiked with (-)-ephedrine (2), dimethinden (3), and propranolol (4)
A) 20 mM citric acid buffer pH 2.5 with 2 % (w/v) carboxymethylated β-CD
B) 1.5 % (w/v) carboxymethylated β-CD adjusted with Tris to pH 5.5

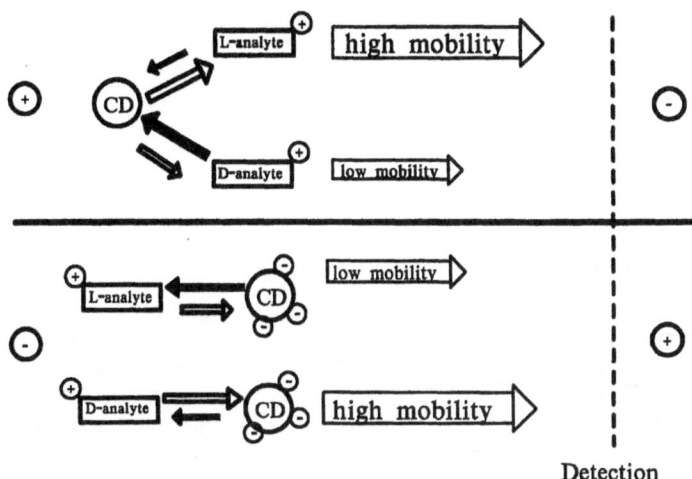

Detection

Fig. 7-23 Schematic representation of the reversal of the migration sequence of enantiomers

Large changes occur in the selectivity of the enantiomeric separation and the migration sequence as a result of the pH change. By converting the uncharged cyclodextrin into the anionic form, the migration sequence of the cationic enantiomer pair can be reversed. This can be seen for ephedrine (peak 2) in this figure. The reversal of the migration sequence of the enantiomers is illustrated schematically in Fig. 7-23. If the D-analyte has a higher affinity for cyclodextrin in its uncharged form, its mobility is decreased more in comparison to the L-analyte. Since the charge of the analyte establishes the migration direction, the D-analyte reaches the detector after the L-analyte and is therefore the slower (Fig. 7-23, upper). If an ion-pair is formed at higher pH values by the same solute, the charge of the cyclodextrin determines the migration direction for cationic enantiomers as well. In this case the D-enantiomer with the highest affinity arrives first at the detector (Fig. 7-23, lower).

In the separation of neutral solutes with charged cyclodextrins, additional problems may arise if the solutes have low solubility in the aqueous system without CD. Often the sample solubility is improved by the addition of CD to the sample vial. If a higher CD concentration is thus used than is present in the buffer, undesirable overloading effects may appear (similar effects occur when high concentrations of organic solvents are used for the same purpose). Fig. 7-24 shows the results when hydrodynamic injection was carried out from such a solution. In case A, a resolution is no longer recognizable, in case B the sample solution was diluted 1:10. A separation with much band broadening is recognizable. One can elegantly obtain a highly efficient separation from solution A if the sample is injected electrokinetically. Only the charged cyclodextrin complexes migrate into the capillary and the system is thus not overloaded (case C).

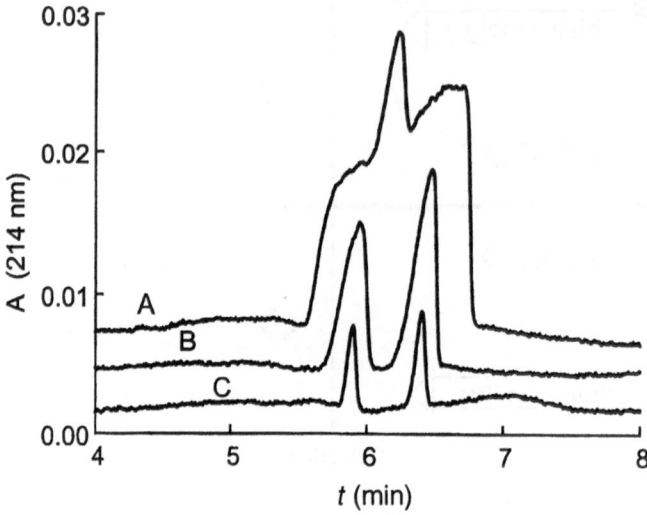

Fig. 7-24
Effect of mass over-
loading on resolution
and peak shape of a
neutral analyte in the
presence of anionic cy-
clodextrins

Separation conditions: L = 20/27 cm, E = 370 cm, 1 % (w/v) carboxymethylated β-CD (CAM), T = 20°C, capillary coated with 4 % T linear polyacrylamide, sample: oxayolidinone derivative dissolved in an excess of CAM

7.2 Other Separation Systems

The chiral selectors presented above are often not used alone, but together with other buffer additives. Micelle-forming detergents such as SDS that are used most frequently, form a second separation system in addition to the chiral selector. The most used combinations will be discussed briefly.

Enantiomeric separations can be optimized by adding micelle-forming components. As was just shown, the selectivity of cyclodextrin-containing systems can be frequently improved by the addition of SDS. In this case, the micellar system provides for the separation of the individual sample components and the cyclodextrin as chiral selector enables their separation into the pure enantiomers. However, a negative influence is often observed from a combination of detergents and cyclodextrins. Detergents with long alkyl chains can imbed themselves inside the cyclodextrin cavity and thereby hinder the chiral recognition of the solute.

The selectivity of such a system can be improved by the addition of another chiral selector. For example, for the combination of "chiral crown ether plus cyclodextrin" a synergistic effect was achieved for a few samples. The use of two different cyclodextrins in the same buffer system can sometimes also lead to improved selectivity. In general, however, the introduction of a second selector and thus a second equilibrium system results in a loss of selectivity.

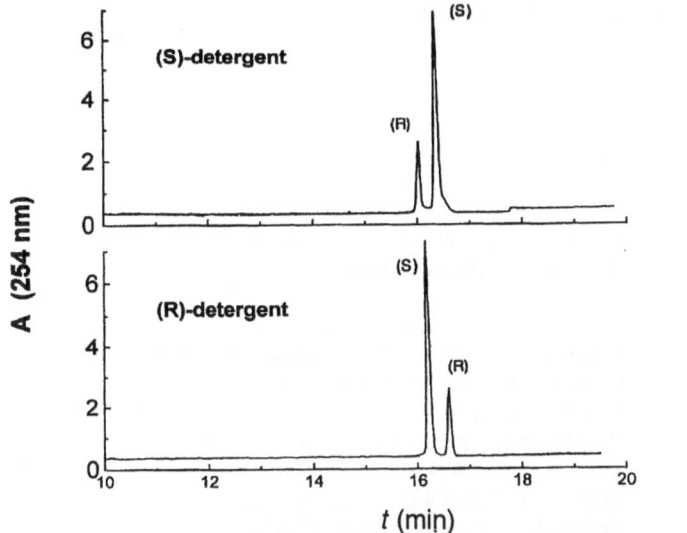

Fig. 7-25
Enantiomer separation
with chiral detergents
[94]

Reversal of the migration sequence of enantiomers by chiral MEKC.
Separation conditions: L = 60 cm, 50 μm i.d., detection at 254 nm, separation potential 15 kV, 50 mM phosphate buffer pH 7.0 with 25 mM (R) or (S)-dodecoxy-carbonyl-valine, sample: 3:1 ratio of (S)- to (R)-benzoin

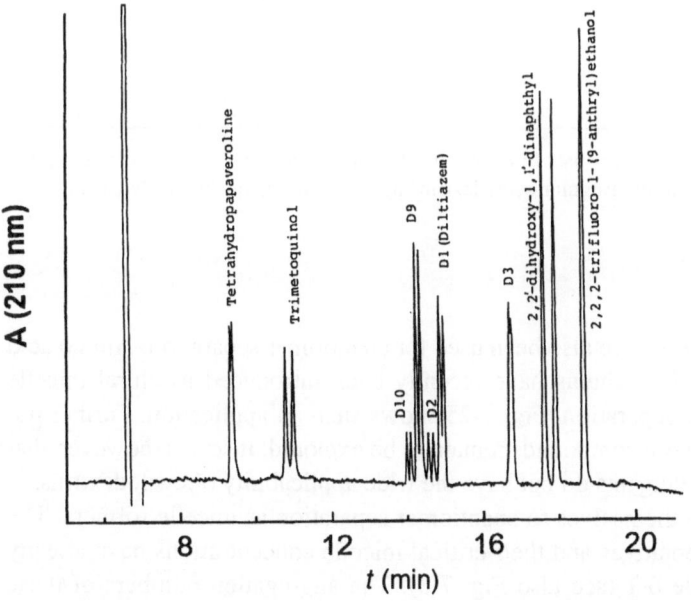

Fig. 7-26
Enantiomer separa-
tion of Diltiazem
analogs with bile
acids as micelle
formers [100]

Separation conditions: L = 50/65 cm, 50 μm i.d., separation potential: 20 kV, buffer: 50 mM sodium taurodeoxycholate in 20 mM phosphate-borate buffer pH 7.0, detection at 210 nm, room temperature

Table 7-4 Sources of important derivatized cyclodextrins

Cyclodextrin	Sources
Methylated CD	- Wacker Chemie, Division L, Biotechnology, Hanns-Seidel-Platz 4, D-81737 Munich, Germany - Cyclolab, P.O. Box 435, H-1525 Budapest; 1026 Budapest, Pusztaszeri ut 59-67, Hungary (see also chemicals catalogs)
Hydroxypropylated CD	- Wacker Chemie, Division L, Biotechnology, Hanns-Seidel-Platz 4, D-81737 Munich, Germany - Roquette Frères, 4 Rue Patou, F-590022 Cedex, France - Janssen, Biotech N V., Lammerdries 55, 2430 Olen, Belgium Distributed by Spectrum Chemical Mfg. Corp. 755 Jersey Ave, New Brunswick, NJ, USA - Cyclolab, P.O. Box 435, H-1525 Budapest; 1026 Budapest, Pusztaszeri ut 59-67, Hungary
Sulfobutyl ether	- Isco, Inc., 4700 Superior St., P.O. Box 5347, Lincoln, Nebraska 68504, USA
Carboxymethylated CD, Carboxyethylated CD, Carboxymethylated CD polymers Succinylated CD Phosphated CD	- Cyclolab, P.O. Box 435, H-1525 Budapest; 1026 Budapest, Pusztaszeri ut 59-67, Hungary

To our knowledge, the cyclodextrins listed in Table 7-2 but not mentioned in this Table are up to now not available commercially. Regular cyclodextrins are listed in many chemicals catalogs.

Chiral micelle formers have also been used for enantiomer separation. Amino acid derivatives with long alkyl chains have recently been introduced as chiral micelle formers for enantiomer separation. Fig. 7-25 shows such an application. Further potential applications of these compounds remain to be explored. It seems, however, that these optically active detergents do not have the wide applicability of cyclodextrins.

Bile acids also lend themselves to enantiomer separation as micelle formers. The most important representatives and their critical micelle concentrations have already been presented in Table 6-1 (see also Fig. 7-3). The aggregation numbers of these micelles are conspicuously low, suggesting that they rather resemble a sandwich structure. Therefore they are suited for the enantiomeric separation of relatively flat and rigid components. This was confirmed in practice, e.g., via the high achievable selectivities in the separation of binaphthol and Diltiazem derivatives (Fig. 7-26). In general, the same fundamental rules apply to the optimization as were discussed for MEKC.

Since only few derivatized cyclodextrins can be found in the usual laboratory chemical catalogs, some important sources are summarized in Table 7-4.

Capillary electrophoresis has proven to be a very versatile analytical method for enantiomeric separations. With a few cyclodextrins that are available in the laboratory and some standard buffers, positive results can be readily observed. Due to the high efficiency, enantiomeric separations become perceptible even when extremely small selectivities are achieved. Further optimization steps soon lead to baseline separations. Method development appears much easier and faster than in LC, where a variety of expensive chiral stationary phases must be kept in stock and various eluent combinations have to be tested to achieve a chiral separation.

8 CAPILLARY GEL ELECTROPHORESIS (CGE)

The tremendous rise in capillary electrophoresis, especially in capillary gel electrophoresis (CGE), was also, in part, induced by the American Genome Project. Initially, DNA molecules from sequencing reactions or restriction fragments were separated almost exclusively by CGE [112, 113]. Of the gel types used in classical slab gel electrophoresis, primarily acrylamide, agarose, and cellulose have been employed as a matrix in capillaries. These gels differ greatly in their physical properties, the most important being the viscosity, the stability in an electric field, and the pore structure and size.

The utilization of gels in electrophoresis is based on the fact that biopolymers, as polyanions or polycations, exist with the same surface/charge ratio and cannot be separated in a normal electric field without auxiliary aids. Since these biopolymers differ greatly in size, however, a gel can be used to influence the mobility of a large polymer more than that of a smaller one. This then leads to a separation according to molecular size, i.e., according to increasing molecular weight, as presented schematically in Fig. 8-1.

The principal fields of application of gel electrophoresis are the separations of DNA molecules and of proteins denatured by SDS. Otherwise, gels are used pre-

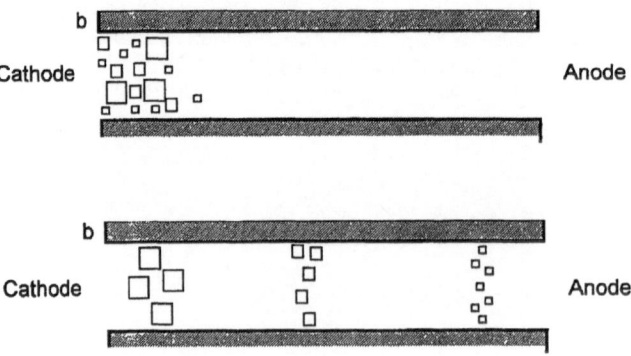

Fig. 8-1 Schematic repreentation of the separation of ionic biopolymers on the basis of molecular size

dominantly in classical electrophoresis as stabilizing gels which do not contribute to separation.

In the following, the particular types of gels will be discussed and their primary fields of application in capillary electrophoresis will be demonstrated.

8.1 Acrylamide-based Gels

In general, distinction is made between gels that display a certain degree of crosslinking (crosslinked gels, consisting of two monomer units) and those that are built up from one monomer unit and are designated as linear gels. These gels are built up from a network of linear polymer chains and their cohesion is based on physical interactions ("physical gels"). In contrast to this are the crosslinked gels in which the individual polymer chains are crosslinked with each other and are therefore substantially more rigid in their physical behavior ("chemical gels" because there are covalent bonds between the strands). Furthermore, they possess a substantially more definite pore size which greatly affects the separation efficiency of the gel.

The structure of a crosslinked polyacrylamide gel (PAA) is shown in Fig.8-2. It is prepared by copolymerization of acrylamide with bisacrylamide as in slab gel electrophoresis. The total monomer content is given as % T (g of monomer/100 mL buffer solution) and the proportion of the crosslinker is designated as % C (percent crosslinker of the total monomer content). The higher its content, the more crosslinked is the gel, and the higher is its viscosity. The migration resistance increases in the same direction. These gels must be polymerized in the capillary with the required buffer. A buffer change afterwards is difficult and time-consuming to achieve.

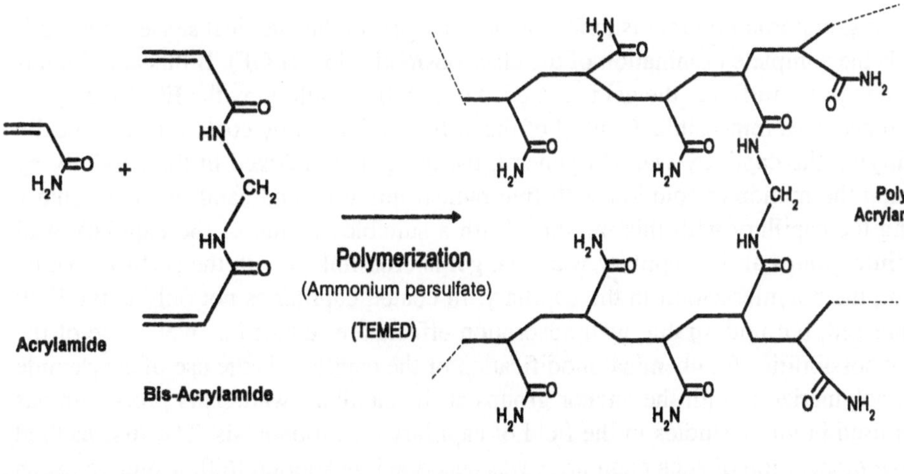

Fig. 8-2 Structure of the monomers acrylamide gels with a section of a polymer

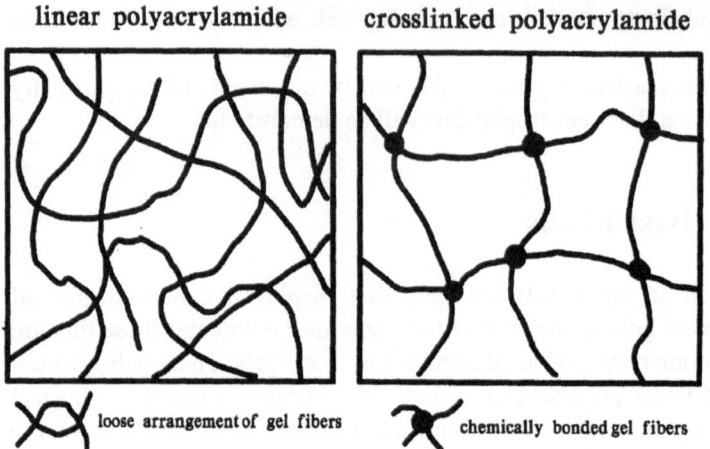

Fig. 8-3 Structural schemes of a linear polyacrylamide and a crosslinked polyacrylamide

Linear polyacrylamides (LPA) [114, 115] are prepared without a crosslinking agent and differ greatly in their properties from the crosslinked gels, even if they agree in their total monomer concentration. These gels are liquid, can be pressed hydrodynamically into the capillary and exchanged after each separation. Fig. 8-3 shows the structural differences of both gels. A 3 % T "gel" without a crosslinking agent is a low-viscosity liquid (therefore also called "liquid gel"), whereas even with only 5 % C in the total monomer content a gelatinous crosslinked gel is obtained.

8.1.1 Preparation and manipulation of gel-filled capillaries

The most important prerequisite for the use of a gel (in the classical sense) in a capillary is the complete elimination of the electroosmotic flow (EOF). If this condition is not adequately fulfilled, the gel is extruded out of the capillary by the EOF and separation becomes impossible. Control of the EOF is achieved by coating or chemically modifying the capillary wall. In general, the gel is polymerized in the capillary by treating the monomer solution with free radical initiators and catalysts, and rapidly filling the capillary with this mixture. With a suitable coating of the capillary wall (olefinic groups at the capillary wall) the gel is crosslinked with the surface coating during the polymerization in the capillary. In coated capillaries not only is the EOF suppressed, but undesirable wall adsorption effects are reduced as well. One of the many possibilities for chemical modification of the capillary is the use of acrylamide to copolymerize it with the anchor groups at the capillary wall. This procedure has been used in many studies in the field of capillary electrophoresis. The first method for the preparation of such capillaries was described by Hjertén [60]. Moreover, such capillaries coated with linear polyacrylamide represent at present the most widespread type of coated capillaries in CE and are also commercially available from some of the manufacturers of CE instruments. A modified procedure for the coating of capillaries and preparation of linear polyacrylamide gels is presented below:

1) Pretreatment of the capillary:
   The pretreatment with alkalies and acids is the first step in the coating process. First, flushing with 1 M NaOH (5 h), is followed by distilled water (10 min) and then by 0.1 M HCl (30 min); the acid is rinsed out completely with distilled water (test with pH paper).

2) First step of capillary coating:
   Water is removed with methanol from the capillary. The latter is then stored for at least 12 hours filled with 50 % (v/v) methacryloxypropyltrimethoxysilane in methanol, with the ends of the capillary closed. The capillary is then flushed with methanol, followed by a mixture of methanol and water, and, finally, with distilled water.

3) Preparation of the acrylamide monomer solution:
   According to the gel concentration desired in the later capillary filling, a certain monomer concentration of acrylamide must be present in the buffer. In this example, a 4 % T linear polyacrylamide gel is to be polymerized onto the capillary wall. Since the polymerization occurs in the capillary, the gel must be removable under pressure after the polymerization. In this respect, monomer concentrations in excess of 6 % T cause difficulties. Prior to the polymerization, the prepared solution must be degassed at 0.1 atm. for 2 hours (3 mL of prepared solution) because oxygen acts as a polymerization inhibitor.

4) Second step of capillary coating: polymerization of the gel in the capillary:
   The degassed monomer solution thus prepared is treated with 5 µL of a 10 % (w/v) ammonium persulfate solution and 5 µL of tetramethylethylenediamine (TEMED) based on 1 ml of monomer solution and the prepared capillary is immediately filled with this solution; the ends of the capillary are closed off with plastic stoppers without delay. The solution can be filled with a µL-syringe that is connected via a Teflon adapter to the capillary. In filling the capillary with the monomer solution, care must be taken not to let air bubbles into the capillary. The polymerization in the capillary requires about 12 hours.

The gels thus prepared in the capillary may also be used for the separation of DNA fragments by capillary gel electrophoresis. The gel prepared without a crosslinking agent can be removed from the capillary and replaced with other gel matrices of higher or lower viscosity. If the gel is removed from the capillary, its wall retains a layer of linear polyacrylamide that greatly suppresses the EOF. These capillaries can be used as coated capillaries in the open tube mode with the popular buffer systems. Gels prepared with a crosslinking agent (>1%C) cannot be removed from the capillary because this destroys the gel matrix.

Capillaries prepared in this way show no measurable EOF for longer periods of time below pH 8. Their utility above pH 9 is limited to a few days, however.

Coated or filled capillaries are difficult to handle; hence some generally to be observed precautionary measures are described below:

Handling of coated and gel filled capillaries:

- A detection window cannot be burned into coated or gel-filled capillaries. This should be done before the treatment of the capillary. The polyimide coating can also be removed by careful scraping with a razor blade, although this does require some skill.
- The cutting of capillaries to the proper length should also be performed with care. After scratching the capillary at the intended break point, the excess piece should be removed at a right angle to the capillary so that no gel is pulled out of the capillary.
- Prior to the first use of the capillary, the full separation potential should not be applied immediately. It is better to apply a low potential initially, and then increase it stepwise until the actual separation potential is reached. This process can be completed in less than 20 min. A constant current and a constant baseline show the conditions have equilibrated. The observed courses of the base-

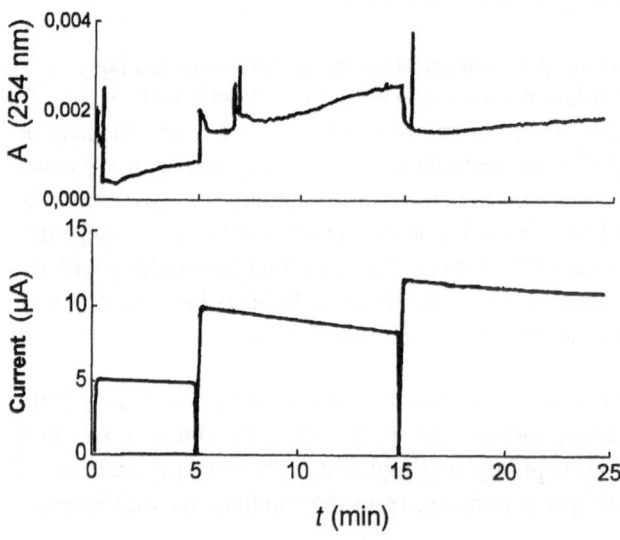

Fig. 8-4
Plot of the baseline and current of a freshly prepared gel capillary based on crosslinked acrylamide

line and current during this equilibration is depicted in Fig. 8-4 for a newly prepared capillary. The removal of polymer residue can also be observed in the detector.

– For longer usage of gel-filled capillaries, care should be taken to store the ends
  of the capillary in a wetting liquid (buffer), in order to prevent the gel from drying out at this site. This would destroy the capillary. For a linear polyacrylamide
  coating the buffer should be rinsed out with distilled water, after which the capillary can be dried in air.

8.1.2 Crosslinked polyacrylamide gels

The capillaries prepared as described above are superior because of the high integrity
of the capillary wall coating in separation systems. Very high efficiencies are
achieved because the gel and capillary wall have the same chemical composition. Up
to now, the highest efficiencies in CE (30 million theoretical plates per meter) were
achieved with crosslinked gels covalently bonded to the capillary wall, which had
been prepared by this method.

The selectivity of the gels can be varied by means of the ratio of monomer concentration used (% T) to the concentration of the crosslinker (% C). Table 8-1 summarizes the properties of the crosslinked, polyacrylamide-based gels.

Table 8-1 Overview of the areas of application of crosslinked polyacrylamide gels

Crosslinked polyacrylamide gels (acrylamide-bisacrylamide gels)		
Monomer concentration	Properties	Areas of application
Various proportions of %T and %C in the controlling the selectivity	solid gel, non-flowing and not exchangeable; temperature sensitive; bonded to the capillary; requires high integrity of capillary coating	- DNA sequencing reactions - oligonucleotide analysis - hyaluronic acid, oligo-saccharides - SDS-protein complexes

%T = g total monomer content (w/v)
%C = (g crosslinker/total monomer content)   100

In practice, however, the fixed gels do exhibit some disadvantages:
- Through inappropriate handling of these gel-filled capillaries (insufficient wetting of the capillary ends with buffer) the gels at the capillary ends may dry out and become unsuitable for further use. This may occur even during an exchange of capillaries in the instrument.
- The buffer medium in the capillaries could not be exchanged or was very time-consuming.
- The thermal stability of such types of gels was inadequate. Dissolved gas in the buffer led at higher temperatures to bubble formation in the capillary. The same effect was also observed at high field strengths (>500 V/cm). The formation of an air bubble in a gel-filled capillary always leads to local destruction of the gel and makes the capillary useless.
- Since the same gel is always used for separation, aging effects through bleeding of the gel are unavoidable. In comparison to this, in classical gel electrophoresis the gel was used for only one analysis. This does not seem reasonable for modern CGE.
- An exchange of such gels in the capillary is hardly possible owing to the covalent bonding to the wall.
- The preparation of one's own capillaries required much know-how in the modification of the surface, especially for conducting the polymerization in the capillary.

8.1.3 Linear polyacrylamide gels (LPA)

The problems described above can be largely avoided if uncrosslinked gels are used which can be simply exchanged after every separation.

For these gels the outer appearance depends exclusively on the monomer concentration. Table 8-2 summarizes the properties and main areas of applications of uncrosslinked, linear polyacrylamide gels. At low monomer concentrations of acrylamide one can no longer speak of a gel in the original sense because up to about 4 % T only an easily flowing high-molecular weight polymer solution is present. These high-molecular weight polymer solutions are also designated as liquid gels, polymer matrices, or sieving buffers. This polymer solution is introduced by pressure into the capillary; it is also present in the buffer containers. This enables the liquid gels and the buffer in the capillary to be exchanged very simply after each analysis. Thus, for each new separation a fresh, unused separation matrix of identical properties is available. This is also reflected in the stability of the method and in the analytical results. Bubble formation, contaminated samples, and dehydration of the capillary ends pose no problems. The stability of such a capillary can be gathered from Fig. 8-5, which shows that after 400 analyses the capillary was still intact.

During the first 50 runs the separation matrix was replaced only after every 9 runs and it is apparent that the migration times of the fragments separated fluctuate consid-

Table 8-2 Range of application and properties of linear polyacrylamide gels of different concentrations

Pure acrylamide gels (linear, non-crosslinked polyacrylamide gels)		
Monomer concentration	Properties	Range of applications
0-6 % T (% w/v)	Liquid ("liquid gel"), exchangeable after each analysis	-DNA restriction fragments -DNA sequence analysis
6-9 % T (% w/v)	Highly viscous (still flowing), exchangeable after each analysis under high pressure	-DNA restriction -DNA sequence analysis -oligonucleotides -SDS protein complexes
9->12 % T (% w/v)	Gelatine-like (solid gel, non-flowing)	-DNA sequence analysis -oligonucleotides -SDS protein complexes

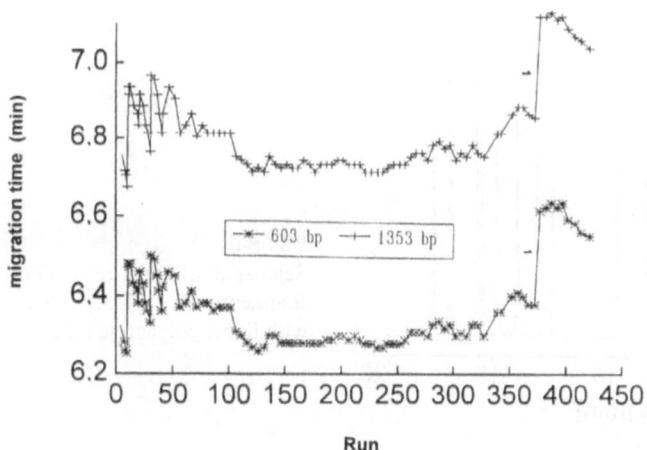

Fig. 8-5 Stability test of a capillary filled with a linear polyacrylamide

Separation conditions: L = 30/37 cm, E = 300 Vcm⁻¹, T = 30 C, 3%T, 0%C LPA, 0.1 M TBE pH 8.3, sample: Phi X 174 RF DNA Hae III; 1. 420. analyses, representative plot of the migration times of the 603 and 1353 base pair fragments

erably. If the LPA solution is replaced after each separation by a flushing procedure (LPA is also present in the buffer container), very constant migration times are obtained (runs 60-350). After 350 analyses a sudden loss in separation efficiency for the longer base pair fragments was established. The general question was whether the separation matrix had aged or the capillary wall coating had been destroyed. By utilizing a newly prepared LPA solution the original separation could be restored. Since the temperature was not controlled during the polymerization, different migration times were obtained for the DNA mixture. This shows, moreover, that the migration time varies from batch to batch and depends greatly on the polymerization conditions.

To attain the highest possible efficiencies and selectivities, the EOF should be eliminated. With flow inside the capillary, the separation efficiency diminishes. It must be mentioned here, however, that with these and other gels it is possible to operate without coating the capillary. The resolution is generally poorer and the separation starts with the highest molecular weights since the slowest anions are detected first after the EOF. In coated capillaries separation occurs according to increasing chain lengths of the DNA fragments.

8.1.3.1 Separation of DNA fragments with LPA

The main area of application of LPA is the analysis of DNA fragments. They can also be used for DNA sequencing. Fig. 8-6 shows the separation of restriction fragments

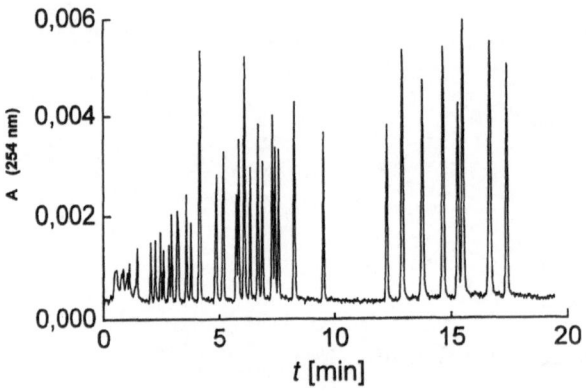

Fig. 8-6
Separation of DNA restriction fragments in a capillary filled with linear polyacrylamide

Separation conditions: L = 40/47 cm, E = 150 Vcm^{-1}, T = 50 C, buffer: 0.1 M TBE pH 8.3 with 3%T, 0%C LPA, detection: 254 nm, sample: pBR 322 MSP I, pBR 322 Hae III (peak assignment from right to left: 622 base pairs (Bp), 582 Bp, 540 Bp, 527 Bp, 504 Bp, 458 Bp, 434 Bp, 404 Bp, 309 Bp, 267 Bp, 242 Bp, 238 Bp, 234 Bp, 217 Bp, 213 Bp, 201 Bp, 192+190 Bp, 184 Bp, 180 Bp, 160+160 Bp, 147+147 Bp, 123+123+124 Bp, 110 Bp, 104 Bp, 89+90 Bp, 80 Bp, 76 Bp, 67 Bp, 64 Bp, 57 Bp, 51 Bp, 34+34 Bp) with a capillary filled with "linear gel" (3 % T, 0 % C). Since all fragments are present in an equimolar ratio to each other, but the larger fragments have more absorption units (nucleotides) in the molecule, the peak area also increases with chain length. As a result, the short fragments appear as small peaks and because of their low migration resistance in the gel, they are first in the electropherogram. The DNA fragments were numbered assuming increasing molecular size. The number of theoretical plates amounts to about 600,000 to 1 million plates/m. At these high plate numbers it is desirable to operate the capillary in an extended, linear configuration, as coiling the capillary may lead to loss in efficiency. However, this loss in efficiency due to the capillary configuration has only been observed at very high efficiencies in CE (>1 million theoretical plates).

It has been found that the temperature also exerts a large influence on the selectivity and efficiency. In general, efficiencies decrease with increasing temperature, although in some fragment ranges greater resolution can be achieved due to improved selectivity. The effect of temperature, field strength, pore structure of the gel, and buffer additives on selectivity and the migration sequence of the analytes will be addressed in the discussion on migration models.

At higher monomer concentrations the polymer solutions become more and more gel-like and highly viscous, so that beyond a concentration of 8 % T they must be po-

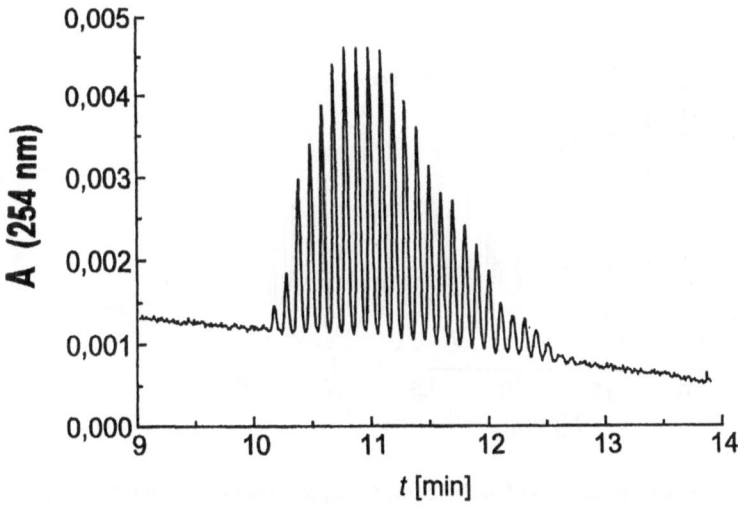

Fig. 8-7 Separation of a polydeoxyadenosine (pd(A) 40-60) oligonucleotide mixture in a 12%T, 0%C gel-filled capillary

Separation conditions: L = 30/37 cm, E = 500 Vcm^{-1}, buffer: 0.1 M TBE 7 M urea pH 8.3, T = 50°C

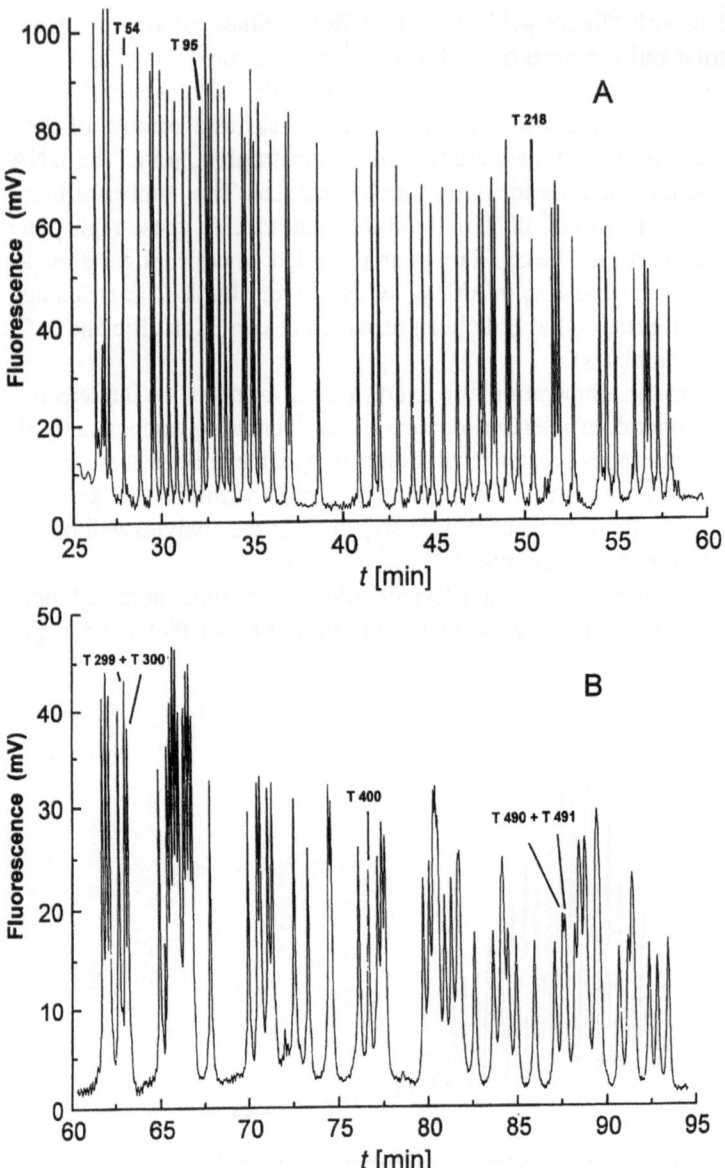

Fig. 8-8 (A+B) DNA sequencing reaction of an M13mp18 template with ddTTP termination [116]

Separation conditions: L = 30/45 cm, E = 200 Vcm^{-1}, T = 25 C, buffer: 100 mM Tris, 100 mM boric acid, 2 mM EDTA, 4 % T linear polyacrylamide (temperature during polymerization = 0°C), 30 % (w/v) formamide, 3.5 M urea, detection PMT (interference filter 520 nm), argon ion laser 488 nm, (to highlight the sequencing efficiency of these gels, some bases of this thymine sequencing are numbered.)

lymerized in the capillary because they cannot be introduced into it with the pressure conditions prevailing in current instruments. The selectivity of these high molecular "linear gels" (e.g. 12 % T) resembles that of crosslinked gels that can only be polymerized in a capillary. However, since in these gels there are no covalent bonds among the chains nor to the capillary wall, high stress on the gel does not result in its degradation. Since the chains are not fixed in these gels, the polymer has a higher flexibility and therefore greater longevity. A separation of an oligonucleotide standard in a coated capillary filled with 12 % T linear polyacrylamide is presented in Fig. 8-7.

These gels are also used in DNA sequencing, for which laser-induced fluorescence detection is used exclusively. The gels are not transparent below 250 nm. The availability of only very small amounts of sample necessitates high detection sensitivity. The detection limit required is about 10^{-11} mol L^{-1}.

Recently, linear polyacrylamide solutions with a low monomer content have also been used in DNA sequencing. These have the advantages of possibly faster sequencing and a separation matrix of longer stability since it can be replaced after every separation. This results in a larger sample throughput, more constant migration times, and thus a smaller number of missequencings. Fig. 8-8 shows a sequencing reaction with chain termination (ddTTP).

8.1.3.2 SDS PAGE (Polyacrylamide gel electrophoresis) of proteins

Until now, CGE was utilized almost exclusively for the separation of DNA molecules because proteins cannot be detected sensitively in the UV region due to low wavelength (<250 nm) absorption by the polyacrylamide. Interactions of proteins with gels may also play a role, so that up to the present only the successful separation of proteins denatured with SDS in gel-filled capillaries has been described. The different charge distributions of proteins in accordance with their pIs, enable them to be separated in a normal buffer system under non-denaturing conditions. In order to achieve separation according to increasing molecular weight, the proteins must bear the same surface/charge ratio. Hence, separation according to molecular size (molecular weight) in a gel is then possible. Proteins are completely denatured with excess SDS and 2-mercaptoethanol (to break up the disulfide bonds). The resulting protein chains bind a constant amount (1.4 g) of SDS per gram of protein, independent of their size and structure. Since the charge of the protein-detergent complex is governed by the SDS taken up, proteins treated in this manner have the same charge/mass ratio.

As a result, uniform mobility is observed in a buffer medium without "sieving properties". If a gel is used, the measured mobility of the analyte is found to be proportional to its effective ionic radius and thus its molecular weight (polypeptide chain). Since there is a linear relationship between the logarithm of the molecular weight and the migration time, the molecular weight of the protein can be ascertained. Fig. 8-9 presents a separation of SDS protein standards of high molecular weight in a gel-filled capillary.

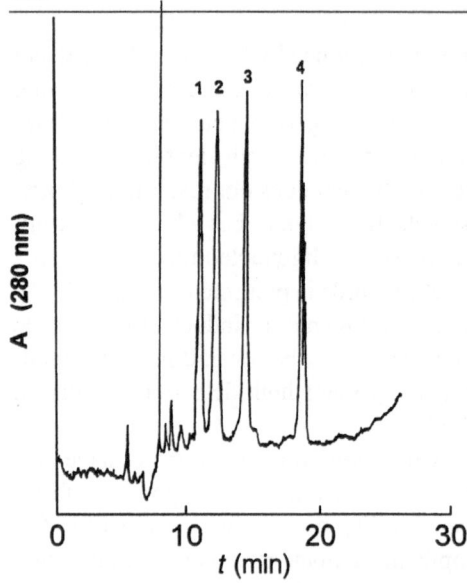

Fig. 8-9.
Separation of high-molecular-weight proteins with crosslinked polyacrylamide gels [117]

Separation conditions: L (effective) = 20 cm, E = 180 V/cm, buffer: 0.12 M Tris, 0.12 M histidine pH 8.8, 0.1 M SDS, 1% 2-propanol, 5 % T, 1 % C polyacrylamide

1 = Ovalbumin
2 = Bovine serum albumin
3 = β-Galactosidase
4 = Myosin

Thus, separation by SDS PAGE in a direction orthogonal to that by IEF constitutes classical 2 D electrophoresis (IEF coupled with SDS-PAGE), a high resolution separation technique.

8.2 Polysaccharide-based Gels and other Polymers

Besides polyacrylamide gels, a series of other water soluble polymers have been used as separation media in capillary gel electrophoresis. Table 8-3 summarizes the polymers used up to the present and their principal areas of application. The chief advantages of these separation matrices lie in their superior UV transparency in the wavelength region below 260 nm. The considerably lower toxicity of these polymers compared to monomeric acrylamide enables simpler manipulation of their solutions. Moreover, these separation polymers also have a low viscosity in the concentration ranges used that permits them to be replaced after each analysis. However, for certain DNA lengths they exhibit lower resolution than the acrylamide gels. Nevertheless, these gels have turned out to be particularly advantageous for the separation of SDS-protein complexes because they allow monitoring at substantially lower wavelengths (214 nm) and therefore with greater sensitivity. Care must be taken to select an appropriate nonabsorbing buffer medium. Fig. 8-10 depicts a separation of SDS-protein standards in a UV transparent dextran solution.

The differences in selectivity between various polymer matrices are based predominantly on their "pore structure". The flexibility of the polymer chains that make

Table 8-3 Summary of the separation polymers other than acrylamides

Other types of gels	Materials used	Applications
Agarose gels	Agarose (wide range)	DNA restriction fragments (up to 1300 base pairs), SDS protein complexes
Cellulose gels	- Hydroxyethyl cellulose (HEC) (SeaPlaque SeaPrep), various concentrations - Methyl cellulose (0.5 %)	DNA restriction fragments (up to 12,000 base pairs) DNA restriction fragments (up to 20,000 base pairs)
Dextrans	Dextrans of various concentrations (molecular weight 10,000 - 2,000,000) ·	SDS protein complexes
Polyethylene glycols	PEG (MW 100,000) of various concentrations	SDS protein complexes

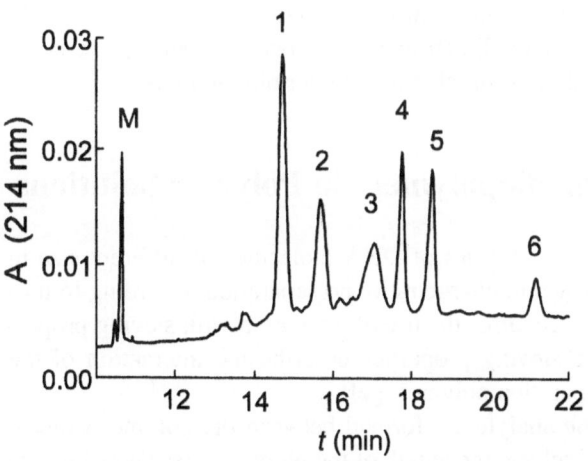

Fig. 8-10 Separation of SDS protein complexes with an exchangeable dextran polymer solution

Separation conditions: L = 30/37 cm, E = 300 Vcm^{-1}, T = 20°C, (Beckman SDS Protein Kit), detection: 214 nm, analytes: Orange G (M), carboanhydrase (1), ovalbumin (2), bovine serum albumin (3), phosphorylase B (4), beta-gallactosidase (5), myosin (6) (see also [117])

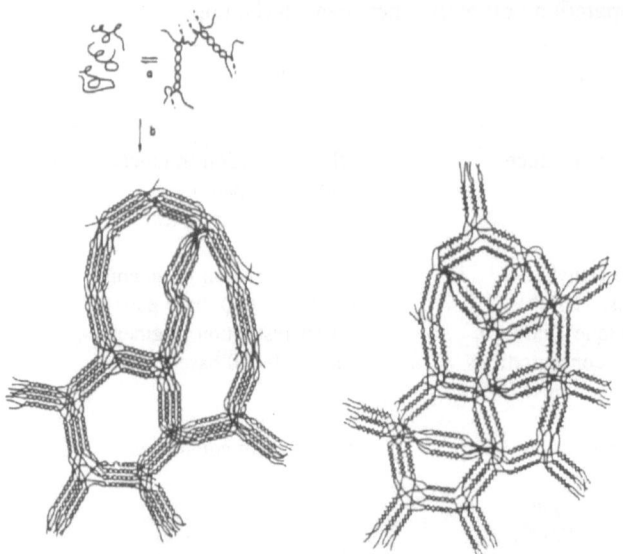

Fig. 8-11
Formation of an agarose
gel and the formation of a
double helix structure.
Left: regular agarose;
right: hydroxyethyl agar-
ose [118]

up the gel exerts considerable influence on its applicability in separations according
to molecular size. For some of these gels the presence of a pore structure has been
discussed. Fig. 8-11 illustrates the formation of such structures, using agarose as an
example. The helical structure of agarose is formed from the monomer strands (left).
In contrast to that, the hydroxyethylated agarose derivative possesses a greatly
changed structure (right). The double helix structure is expanded, causing the pores
to be narrower meshed and the gel more suitable for smaller biopolymers.

8.3 Migration Models of Biopolymers in Polymer Solutions

Since there is no difference in the mobilities of DNA molecules of different size in
free solution due to the same surface to charge ratio, no separation according to molecular size is possible. This, then, requires the use of a medium with sieving properties. In the simplest sense, these sieving properties describe the interaction of the
analyte with the strands of the separation polymer (gel).

Different conformations of the analyte are formed between the polymer strands,
depending on the pore size of the gel and the length of the biopolymers. These conformations are ultimately responsible for the different mobilities and the apparent irregularities. The various conformations of the DNA molecules - such as extended or
relaxed or U-shaped - could be observed by laser fluorescence microscopy (*cf.* Fig. 8-
16). It is evident that a compact or a relaxed conformation develops a different mobility than, for example, the extended form. Similar shapes can be assumed for SDS-protein complexes.

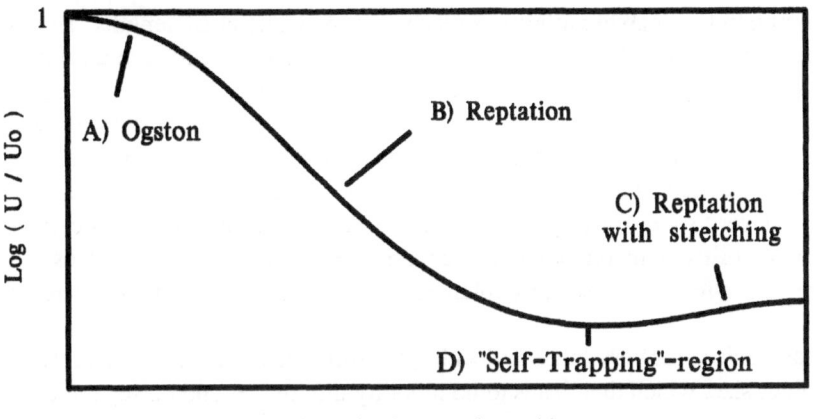

Log (polymer length)

Fig. 8-12 Simulation of the log(analyte mobility) *vs.* log(analyte segment length) at constant pore size of the separation matrix [119]

Different mechanisms are being discussed for the various interactions of DNA molecules. Separation according to molecular size is based on the fact that the larger DNA molecules interact more intensively with the polymer matrix than the smaller ones. Different separation mechanisms may be involved, depending on the conformation and size of the DNA biopolymer and the pore structure of the polymer solution (gel). In general, the mobilities of an analyte in a separation polymer are found to vary with respect to its size in the manner shown in Fig. 8-12 [119]. Four regions can be distinguished in this curve:

A) In the flatter initial portion of the curve (for shorter chains) the mobility changes very little with increasing analyte length, i.e., the selectivity is low for the smaller fragments. In this region, designated as the Ogston region, the pore structure of the gel is larger than the molecular size. Molecules can migrate through the gel without great resistance and retain their globular conformation. This is also the reason for the low selectivity of the analytes. Gels with larger crosslinking or higher gel concentrations must be selected for this separation region.

B) The greatest dependence of the analyte mobility on its chain length and thus also the greatest selectivity is found in the so-called reptation range. In this range, the analyte is larger than the gel pores and can only penetrate them in an extended state via a snake-like movement (reptation). The largest interactions of the analyte with the gel matrix can arise in this way, and is also reflected as a maximum in the selectivity for separation according to molecular weight.

C) A minimum in the analyte mobility is reached at moderate chain lengths: This is ascribed to both ends of the molecule moving in the same direction and becoming strongly entangled with the polymer strands (trapping). This leads to a mobility that does not depend on the molecular size (abnormal migration or even inversion). This trapping does not occur with small molecules, however, as they can quickly free themselves from this conformation. The probability of reaching this state is very small for large molecules. For large fragments trapping can reverse the mobility. This phenomenon remains the focus of scientific discussion because this effect can also be explained by assuming a tertiary structure of DNA.

D) In the last region, for very large molecules, the differences become smaller relative to their size, which then leads to no mobility differences and no separation.

As Fig. 8-13 shows, the dependences measured in practice agree well with the theory. It has, however, been observed that the field strength, temperature, and gel concentration also exert strong influences on the expected migration sequence (small molecules before large ones) and on the mobilities. In extreme cases inversions of the migration sequence have been observed, i.e., a longer DNA fragment migrates faster through the separation gel than a smaller one. This undesirable behavior can generally be eliminated by reducing the applied field strength, raising the temperature, decreasing the gel concentration, incorporating buffer additives such as ethidium bromide as DNA intercalators, and using pulsed electric fields.

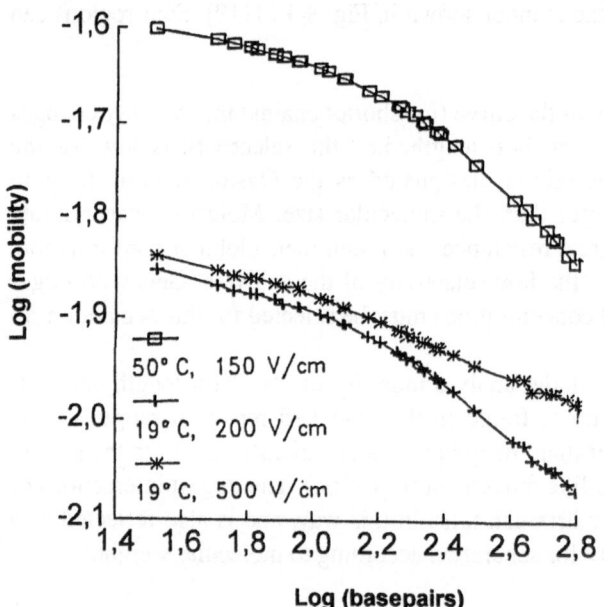

Fig. 8-13
Plot of log(mobility) of a mixture of DNA restriction fragments against the log(number of base pairs)

Separation conditions: L = 40/47 cm, buffer: 0.1 M TBE pH 8.3 with 3 % T, 0 % C LPA, various field strengths and temperatures

The migration behavior of biopolymers has been described quantitatively by the reptation model [120]. The following relationship was found between the mobility of the biopolymer, its fragment length (the number of base pairs in this case), the field strength, the temperature, and the free path length between the polymer strands of the gel:

$$\mu = \frac{Q}{3f}(\frac{1}{B_P} + (\frac{qEa}{3k_b T})^2 + ...$$ (8.1)

where μ is the mobility of the biopolymer, Q the total charge on the molecule, f a frictional factor between the gel strands and the polymer, B_p the number of segments in the biopolymer (here the number of base pairs), q the charge per B_p, E the field strength, a the length of the gel pore, T the temperature, and k_b the Boltzmann constant. It is evident from this equation that the mobility is inversely proportional to the analyte chain length only when the second term of this series is negligible. This holds, however, only when relatively low field strengths (<200 Vcm^{-1}) for CE are used. The relationship between the mobility and the reciprocal of the chain length

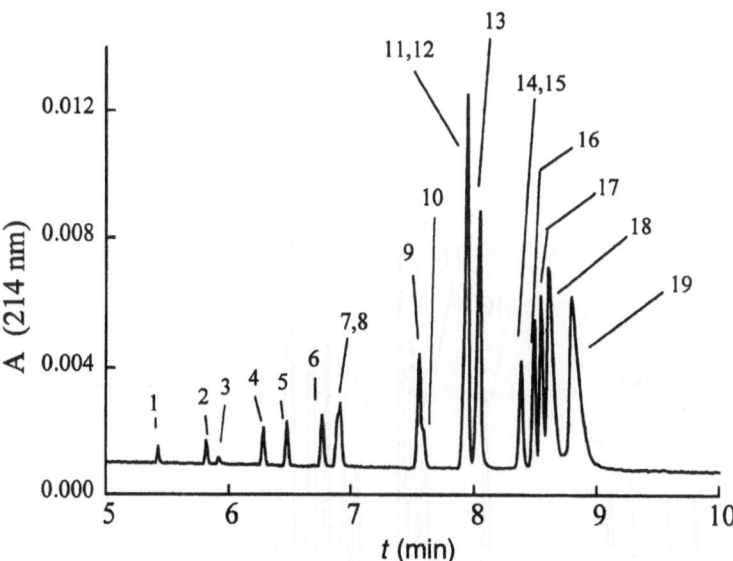

Fig. 8-14 Effect of temperature and field strength in the separation of DNA restriction fragments
Separation conditions: L = 20/27 cm, E = 300 Vcm^{-1}, T = 20° C, 3 % T (LPA), sample: mixture of Phi X Hae III and DNA.
Fragments of λ Hind III:125 (Bp(3); 564(BP(10), 2 027 Bp(14), 2 322 Bp(15), 4 361 (Bp(16), 6 557 Bp(17), 9 416 Bp(18), 23 130 Bp(19).
Fragments of ΦX Hae III: 72 Bp(1), 118 Bp(2), 194 Bp (4), 234 Bp(5), 271 Bp(6), 281 Bp(7), 310 Bp(8), 603 Bp(9), 872 Bp(11), 1078 Bp(12), 1353 Bp(13).

should become clearer at higher temperatures. This was confirmed in practice and is clearly recognizable from Fig. 8-13, where the results from Fig. 8-6 are plotted.

The higher the field strength and the lower the temperatures of the separation system, the less dependent is the mobility on the reciprocal of the chain length. The same effect is illustrated again in Figs. 8-14 and 8-15 by means of two electropherograms. Fig. 8-14 shows a separation performed at low temperature (20°C) and high field strength. The separation at a little higher temperature (35°C) and low field strength is substantially better (Fig. 8-15).

As was already indicated at the beginning of the chapter, different conformations of the analyte exist between the polymer strands, depending on the pore size of the gel and the length of the biopolymers. The conformations are ultimately responsible for the different mobilities and the apparent irregularities. The different conformations reproduced in Fig. 8-16 could be observed by laser fluorescence-microscopy [121]. Many of the observed phenomena and anomalies can be described and explained with the above migration model. It can also explain the above-mentioned reasons for the nonlinear behavior between the migration sequence and the molecular size, and suggest countermeasures to eliminate them.

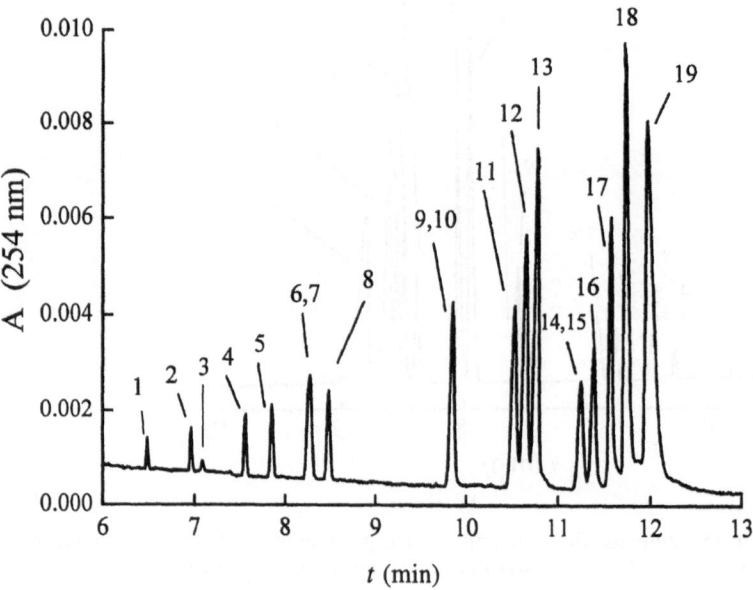

Fig. 8-15 Effect of temperature and field strength in the separation of DNA restriction fragments Separation conditions: L = 20/27 cm, E = 200 V/cm, T = 35° C, 3 % T (LPA), sample: mixture of Phi X Hae III and DNA. Peak designation as in Fig. 8-14.

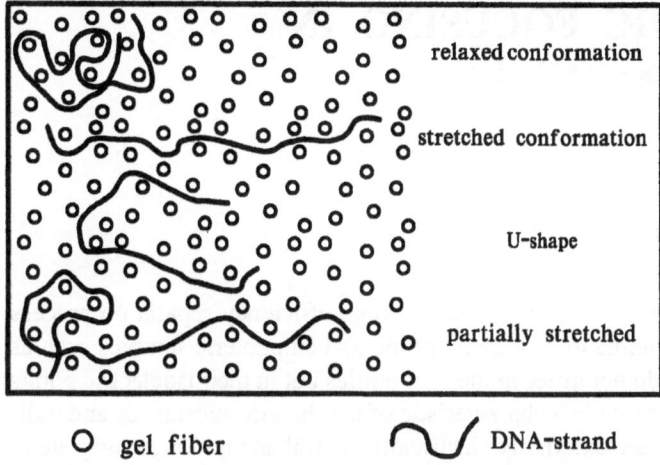

Fig. 8-16 Dependence of the mobility of a biopolymer on its conformation in a separation gel (*cf.* also Fig. 8-12)

9 ISOELECTRIC FOCUSING IN CAPILLARIES (CIEF)

In classical electrophoresis, IEF is an established high efficiency separation process [122-124]. It is used predominantly for zwitterionic and amphoteric samples such as proteins and peptides that do not differ in their mobilities but in their isoelectric points (pI value). The isoelectric point is a characteristic of amphoteric substances and indicates the pH at which they are externally electrically neutral and no longer migrate in an electric field. Thus, at this pH the positive and the negative charges are equal to each other. If an electrophoretic separation according to pI values is desired, a pH gradient is needed that covers the required pI range. Such pH gradients can be generated by placing a solution of amphoteric substances in the capillary whose pI values cover the entire pH range. Aliphatic aminocarboxylic acids with different ratios of amino to carboxylic acid groups are used as ampholytes [125]. Depending on the pH, ampholytes carry a positive or negative formal charge. The "titration curves" for two different ampholytes are displayed in Fig. 9-1. The closer the two pK values are together, the better is the ampholyte. Of course, a series of good ampholytes is required to generate a wide pH gradient.

The pI values of these ampholytes can span the entire or just a small pH range, depending on how large the pH steps need to be selected over the separation distance. Initially, the separation capillary is filled with a mixture of various ampholytes (Fig.

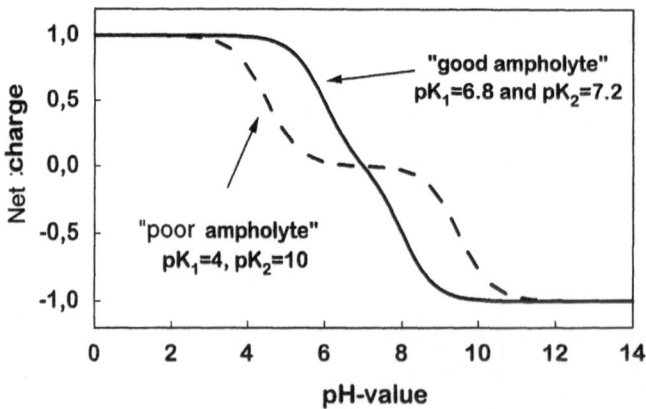

Fig. 9-1
Dependence of the net charge of an ampholyte on pH

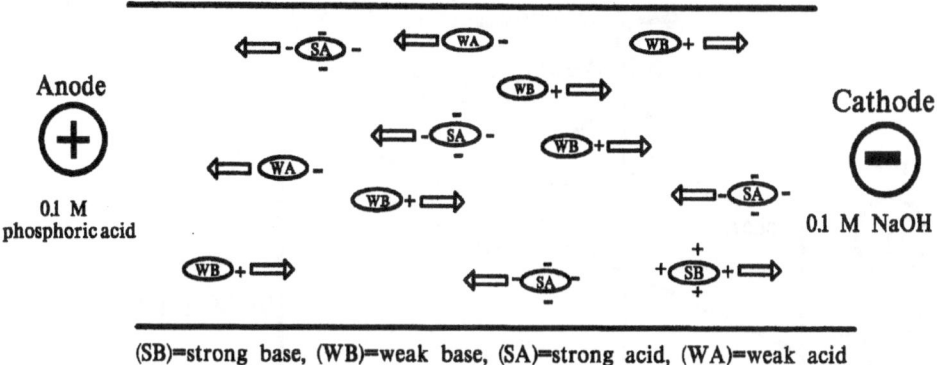

(SB)=strong base, (WB)=weak base, (SA)=strong acid, (WA)=weak acid

Fig. 9-2 Schematic representation of the generation of a pH gradient in IEF

9-2). A strong acid is placed in the anode compartment, a strong base in the cathode vial. After applying the potential, the hÿdrogen and hydroxide ions begin to flow and the ampholytes align themselves according to their pIs. If enough ampholytes with different pI values are present in the mixture, a continuous pH gradient is produced. The formation of a pH gradient under the influence of proton and hydroxide ion migration, the electric field, and the ampholytes is presented in Fig. 9-3. In classical slab gel electrophoresis the pH gradient is immobilized on the gel in order to prevent dispersion by convection currents caused by Joule heating during the passage of current through the electrolyte solution. Proteins migrate in this pH gradient only so long as they possess a net charge. Their electrophoretic migration terminates at the pH corresponding to their isoelectric point. The high efficiencies obtained in IEF stem from the

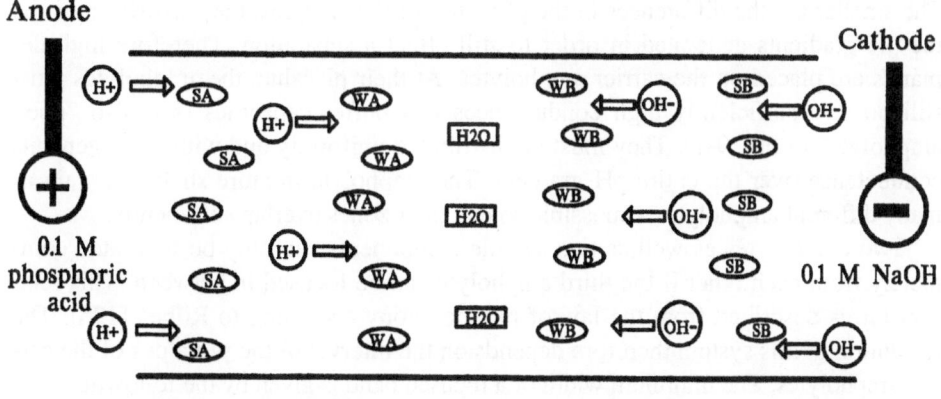

(SB)=strong base, (WB)=weak base, (SA)=strong acid, (WA)=weak acid

Fig. 9-3 Maintenance of a pH gradient in IEF by proton and hydroxide ion flow in an electric field

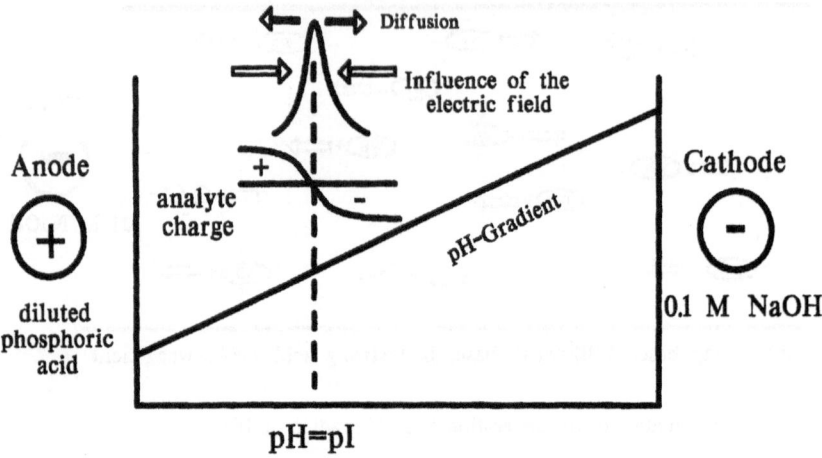

Fig. 9-4 Schematic representation of the focusing properties of a pH gradient in an IEF system

focusing properties of a pH gradient, which allows practically no band broadening through diffusion. This is illustrated schematically in Fig. 9-4. If an analyte molecule attempts to leave the focused band, it acquires a charge in a pH that differs from its pI and is driven back in the opposite direction. Care should be taken that the pH of the anodic buffer is lower than the pI of the most acidic ampholyte in order to prevent migration into the anode compartment, and vice versa for the cathode vial.

The ampholytes can either be added to the buffer or covalently bonded to a gel (immobilized pH gradients (IPG)). The latter variant of IEF leads to very high separation efficiencies because of the very steep gradients that can be achieved with it. The smaller are the differences in the pI values of the analytes, the narrower must be the pH gradients generated in order to still effect a separation. Therefore high demands are placed on the carrier ampholytes. At their pI value, the ampholytes must still possess sufficiently high conductances and buffer capacities (see also "good ampholyte" in Fig. 9-1). They must be distributed uniformly and with homogeneous conductance over the entire pH gradient. The ampholyte mixture should contain as many different ampholytes as possible so that their zones overlap sufficiently.

Two ampholytes as well as two sample components can only be separated completely from each other if the third ampholyte can be focused in between. This relationship is described from the law of pH monotony according to Rilbes [126]. The resolution in this system therefore depends on the interval of the pI values of the carrier ampholytes. The minimum width of a focused band is given by the following:

$$\sigma^2 = \frac{D}{E} \cdot \frac{1}{\dfrac{d(pH)}{dx} \cdot \dfrac{d\mu}{d(pH)}}$$

(9.1)

For an ideal pH gradient, two peaks can be completely resolved only if the difference in their pI values obeys the following relationship.

$$\Delta pI = 3,07 \cdot \sqrt{\frac{D}{E} \cdot \frac{d(pH)/dx}{-d\mu/d(pH)}} \tag{9.2}$$

In transferring IEF to a narrow-bore capillary, stabilizing gels are not absolutely essential. The separation is carried out in free solution, since the capillary itself serves as an anticonvective medium. However, the EOF must be suppressed completely in order to enable formation of the pH gradient. Otherwise the EOF would rapidly flush the ampholyte solution out of the capillary and no focusing would occur. As discussed in Section 3-3, the EOF can be controlled by modifying the capillary surface. The EOF can also be reduced to such an extent that focusing is possible by the addition of highly viscous polymers (e.g., methylcellulose) that raise the viscosity of the buffer solution and thereby dynamically coat the capillary [127]. Dynamic coating has advantages because very high pHs must often be used in IEF and a covalent capillary coating could not long withstand highly alkaline pH values.

Unlike in a slab-gel, in a capillary formation of the pH gradient and the focusing of the proteins occur in the same step. The capillary is filled with the ampholyte solution that already contains the sample. The cathode-end of the capillary is immersed in dilute sodium hydroxide and the anode-end in dilute phosphoric acid. Upon applying a potential, a high current flows initially because the ampholytes are still migrating to their isoelectric points and thus contributing to current flow in the capillary. The proteins migrate simultaneously, as they themselves are ampholytes. The accomplishment of the focusing process is indicated by a decrease in the current to a constant lower value. After focusing, the ampholytes and sample components are distributed across the capillary. In slab gel electrophoresis detection is effected by colorization of the separated zones. In a capillary the analytes must be mobilized again and transported to the detector without loss in the separation. Mobilization can be achieved through an electrokinetic effect. A higher concentration of salt ions is added to the detector side of the buffer: sodium ions at the anode for anode mobilization or chloride ions at the cathode for cathode mobilization. This effects an instability in the pH gradient and the ampholyte and the focused analytes regain their mobility. The time variation of the pH gradient during mobilization is displayed in Fig. 9-5. Mobilization occurs in the direction of the electrode compartment containing the higher salt concentration and the detector must be positioned accordingly.

An alternate means of mobilization involves the application of a pressure difference (pressure or suction) to the system of sample and ampholyte to convey it past the detector. The potential should be maintained, however, to counteract band broadening during the detection process. Because focusing occurs over the entire capillary length, including the section between the detector cell and cathode, proteins with extremely high pI values could escape detection. This can be prevented by filling the capillary with NaOH beyond the detector position.

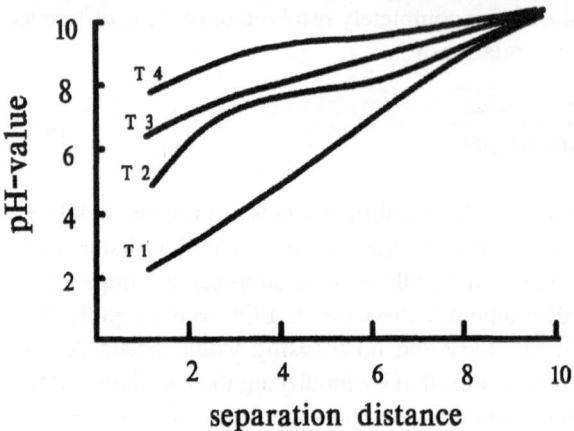

Fig. 9-5
Time variation of the pH gradient during electrokinetic mobilization . The pH gradient dissipates with increasing time (T1-T4)

The advantages of CIEF lie in the ability to use small amounts of sample, the high sensitivity through enrichment in the focused zone, and the extremely good resolution. Since stabilizing gels do not need to be used, interaction between sample and matrix can be excluded. Moreover, the ease of automation and application of high field strengths that yield short analysis times enable high sample throughput. The isoelectric points of many proteins were determined by IEF. An extensive table of pI values of proteins is presented in [128].

Fig. 9-6 shows the application of IEF to low-molecular weight compounds. After a focusing time of 4 min the system was mobilized electrokinetically. This short time is

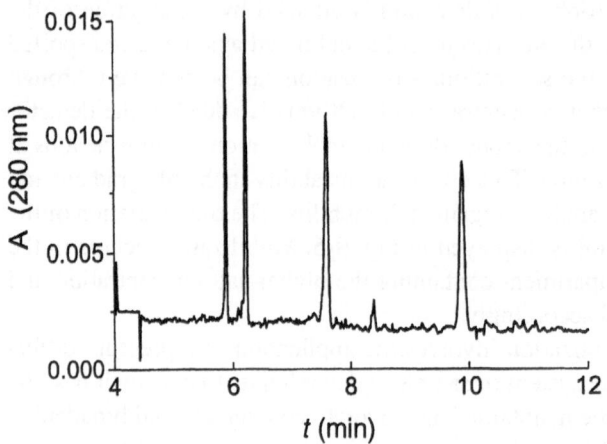

Fig. 9-6
Separation of aminoethylated nitrophenols

Separation conditions: L = 17/25 cm coated capillary (Biorad), Biolyte ampholyte solution (Biorad), focusing: 4 min at 10 kV, mobilization: 10 kV

also adequate for focusing proteins, as shown in Fig. 9-7, using amylase as an example. It should also be mentioned that slab gel IEF requires more than 60 min just for the generation of a pH gradient in the stabilizing gel as only low field strengths can be used. In combination with mass spectrometry, IEF of proteins appears to be becoming a very important separation tool in protein analysis.

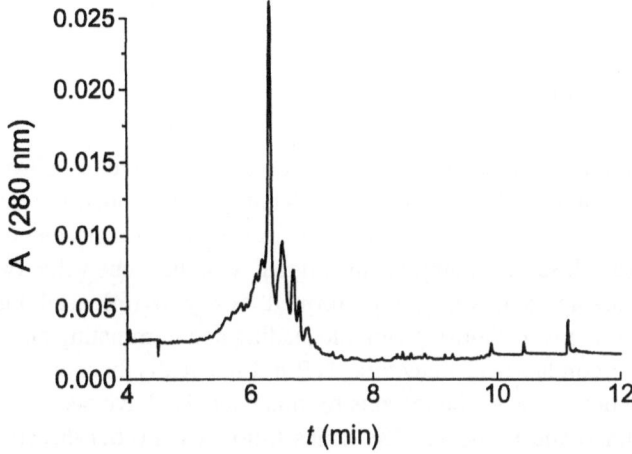

Fig. 9-7 Separation of an alpha amylase standard by CIEF

Separation conditions: L = 17/25 cm coated capillary (Biorad), Biolyte ampholyte solution (Biorad), focusing: 4 min at 10 kV, mobilization: 10 kV, sample: alpha amylase from Bacillus Licheniformis: 1:250 dilution with the ampholyte

10 OTHER SEPARATION TECHNIQUES IN CE

10.1 Isotachophoresis (ITP)

If the electrophoretic separation processes in capillaries correspond to elution chromatography (discontinuous sample application, constant eluent composition, different migration velocities of the sample components) then ITP corresponds to displacement chromatography. In both cases all sample components migrate with the same velocity. ITP was described many years ago. and was then performed mostly in Teflon tubing [129, 130]. But due to the problems in choosing suitable leading and terminating electrolytes and its restriction to conductivity detectors, ITP did not really become accepted as an analytical procedure. One of the reasons for this may also have been that only a step diagram is obtained and no peaks. The zones follow each other directly and are not separated from each other by buffer but each zone contains a single com-

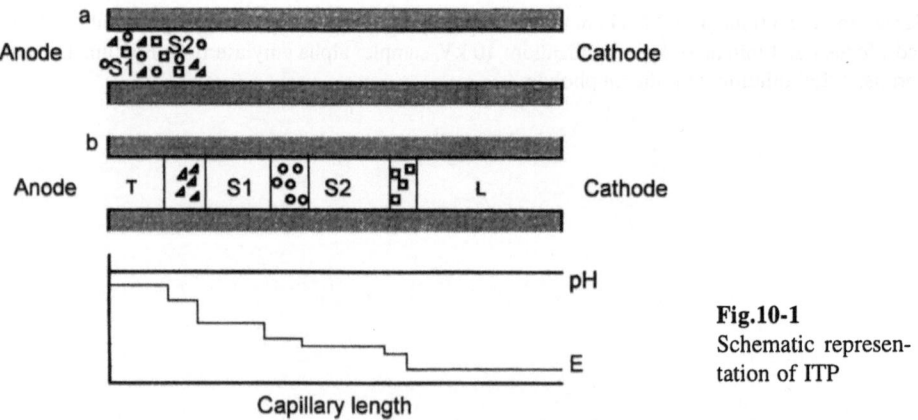

Fig.10-1
Schematic representation of ITP

a) Sample mixture prior to ITP, the capillary is completely filled with the leading electrolyte
b) During ITP after attaining equilibrium: L = leading electrolyte; T = terminating electrolyte; triangles, circles, and squares are the analytes; S1 and S2 are suitable spacers that separate the analyte zones completely from each other.
Lower graph: Variation of the field strength over the separation length and through the analyte zones during ITP

ponent. As in displacement chromatography with stepwise eluent changes, a discontinuous buffer system is used in ITP.

In ITP the sample is injected between two electrolyte solutions with different ion mobilities. The mobilities of the analyte ions must lie between those of the leading and terminating electrolytes. Usually the leading electrolyte has the highest and the end electrolyte the lowest mobility of all migrating ions. After reaching a stationary state, all ions of the same charge migrate at the same rate. This is presented schematically is Fig. 10-1. In each zone in ITP a different field strength prevails, which is constant within each zone; the changes occur jumpwise at the zone boundaries (field jump electrophoresis).

In ITP only ions of the same charge can be separated in a single run. The concentration course of the solute zones can be represented by a rectangular function. The zones follow directly one after the other, as in displacement chromatography. Step signals are obtained with conductivity detection. The step length is a measure of the amount of sample. The arrangement of the analyte between the leading and terminating electrolyte can be demonstrated by UV spectra. Fig. 10-2 shows the extraordinarily sharp zone boundaries between the leading and terminating electrolyte [131]. The

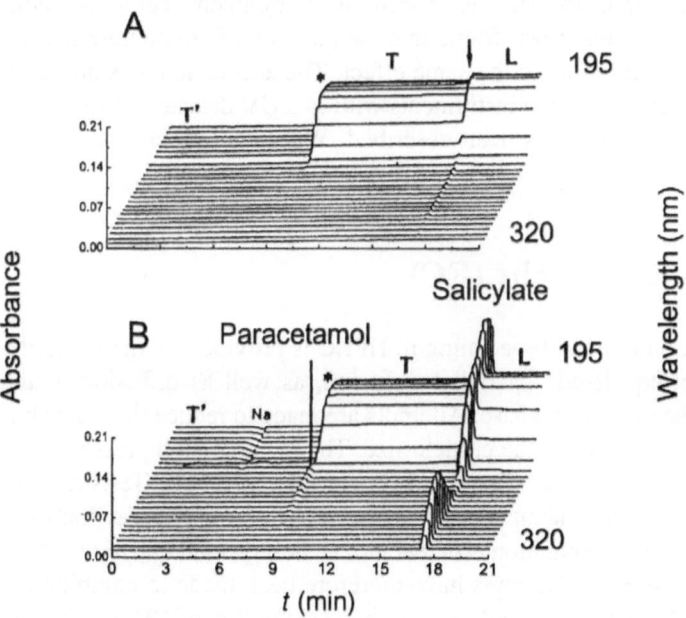

Fig. 10-2 Illustration of the ITP zones with a fast-scanning detector at low wavelengths [131]

Separation conditions:
A) Comparison run (without sample): L (leading electrolyte), T (suitable terminating electrolyte), T' (terminating electrolyte), arrow and star indicate the zone boundaries.
B) Separation of a model sample: Paracetamol (330 µM), salicylate (2.5 mM)

sample is enclosed between the leading and terminating electrolytes, as the example of the salicylate shows here.

ITP is used principally for the separation of inorganic ions and organic carboxylic acids. Owing to detection problems and difficulties in finding appropriate leading and terminating electrolytes for samples of unknown composition, ITP has not yet found wide acceptance. In particular, proteins and other complex mixtures also require suitable spacers, i.e., electrolytes with a migration velocity between that of the analytes in order to separate the zones from each other. The lack of suitable spacers for protein analysis has resulted in but little application of ITP in bioanalysis.

ITP, like displacement chromatography, leads to an enrichment of dilute solutions, and is therefore suitable as a preconcentration step prior to CE separation (*cf.* also Section 4.3.5). It can also overcome the problems of using relatively large volumes of dilute samples. The enrichment step can be carried out directly in the capillary. A short zone of the leading electrolyte (a rapidly migrating organic acid such as acetate [21] is often used for this) is added to a capillary filled with the separation electrolyte, and this is followed by injection of the sample. Sample volumes in the µL range can be used. The usual electrolyte is added as terminating buffer after the sample. Enrichment occurs in the direction of the leading electrolyte. Enrichment factors of about 100 were achieved. For proteins it was found that the addition of ammonium acetate alone to the sample solution produces the same effect. The acetate ion acts here as a leading electrolyte. A coupling of two instruments with two UV detectors, ITP for enrichment and CE for separation, has been described. When the enriched ITP zone reaches the first detector, the system is switched over to CE separation.

10.2 Electrochromatography (EC)

The principal contribution to band broadening in HPLC is provided by the parabolic flow profile that is not equalized by radial diffusion, as well as diffusion of the analytes in and out of the stationary phase. Attempts are made to reduce this contribution to band broadening by reducing the particle size. The required high pressure drop opposes further reduction in particle size below 3 µm. The EOF manifests a plug-shaped flow profile and is independent of particle size of the packing in capillaries in the ideal case. Moreover, high separation efficiencies can be expected due to the plug-shaped flow profile of the EOF. Attempts have therefore been made to combine the high selectivity of HPLC with packed columns with the resolution of CE. The underlying separation principles and stationary phases are those of HPLC, but the eluent (aqueous-organic buffer solution) and the components are transported by electric migration. Very small, nonporous particles (1 µm) are used in order to keep the contribution of the sorption process to band broadening at a low level. Other than theoretical work [132], hardly any real examples of applications have been described in the literature up to the present. The favorable theoretical predictions could also not yet be completely verified experimentally. The reduced plate heights achieved differ only slightly from those attained in HPLC. The detection problem could be one of the reasons for this. In CE detection occurs through the capillary; this is not possible with packed capillaries. Consequently, problems appear that are familiar from micro HPLC relating to the attachment of the detector to the capillary. An example is the separation of some benzene derivatives [133], as presented in Fig. 10-3. Between 40,000 and 60,000 plates/m were achieved with the 28.5 cm long packed capillary. For the 3 μm particle size used this corresponds to a reduced plate height of about 2. This can also be attained in packed columns with a hydrodynamic flow [134].

Further problems arise in achieving the homogeneous packing of capillaries, the maintenance of the packing in the capillary, and the frequent appearance of gas bubbles which still encumber the use of electrochromatography in capillaries. If these problems were solved, substantially higher separation efficiencies could be achieved than in pressure-driven HPLC.

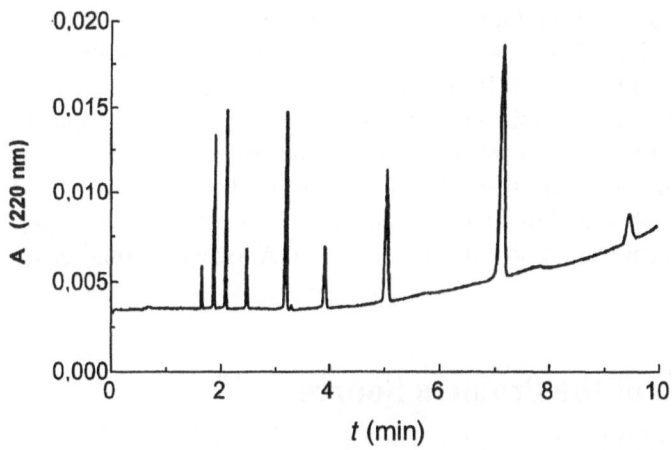

Fig. 10-3 Separation of benzene derivatives by electrochromatography in packed capillaries [133] Separation conditions: L = 28.5 cm, packing material: Hypersil ODS (3 μm), mobile phase: 2 mM tetraborate-80 % acetonitrile, potential: 45 kV (2.0 μA), injection: 2.5 kV, 5 s, EOF velocity: 2.6 mm/s, analytes (left to right): thiourea (N = 46,000, reduced plate height (h) = 2.0), benzyl alcohol (N = 54,000, h = 1.8), benzaldehyde (N = 56,000, h = 1.7), benzene (N = 47,000, h = 2.0), 1,2-dichlorobenzene (N = 54,000, h = 1.8), 1,2,3-trichlorobenzene (N = 52,000, h = 1.8), 1,2,3,4-tetrachlorobenzene (N = 49,000, h = 2.0), pentachlorobenzene (N = 43,000, h = 2.2), hexachlorobenzene (N = 43,000, h = 2.2)

11 A TROUBLESHOOTING GUIDE TO CE

The elimination of problems that lead to instrumental malfunctions during analysis is surely not more difficult in capillary electrophoresis than in other separation methods. Sources of such problems fall into a small number of groups, the most important being in the detector, the capillary surface, the separation buffer, and the sample itself. The troubleshooting strategies discussed below will enable even the CE neophyte to pinpoint the above-mentioned sources of the respective problems quickly and concretely, and then to eliminate them. This chapter is not intended as a substitute for operations manuals for CE instruments, but rather as a description of the more frequently occurring problems and suggestions for solutions. The operations manuals remain the basis for the diagnosis of problems, especially the instrument-specific ones and those associated with the software. Prior to troubleshooting, some simple theoretical considerations should be helpful. If, for example, the capillary length is doubled for the optimization of an analysis while the remaining separation parameters (buffer concentration, pH, potential, ...) are held constant, the analysis time is quadrupled (twice the distance, half the field strength). A peak detected previously after 10 min, will require almost 3/4 of an hour. Consequently, before each troubleshooting prior usage of the instrument should be critically reviewed. A logbook would be an additional aid.

11.1 Determination of the Problem Source

The simplest way to identify the cause of a problem with an instrument or with its operation is for the user to have much experience and a "feel" for potential sources. Another possibility is to check for proper functioning with successive replacement of the individual components of the separation system. However, this method is expensive and time-consuming.

If certain characteristic values of the CE system are known, however, problems can be pinpointed more readily without time-consuming searching or years of experience. Clearly, it is necessary to determine such characteristic values while the instrument is in sound operating condition.

The concept proposed here is based on several years of experience with home-built and commercial CE instruments from seven different manufacturers, during which time many, including some unusual problems had to be detected and rectified. Only in rare cases was it necessary to call a serviceman. If the malfunction is first localized, it is often possible to remedy the problem after a telephone consultation with the service personnel and, if required, to order the appropriate replacement part.

Test 1: Recording a breakthrough curve under flushing pressure

A new 75 μm capillary with an effective length of about 50 cm is flushed and completely filled with water. Its inlet side is then immersed in a solution of 0.5 % (w/w) of benzyl alcohol. This solution is forced through the capillary with the flushing pressure and the detector signal monitored at 214 nm. The height of the step signal later provides information on the detection. The elapsed time until the curve rises permits conclusions to be drawn later concerning clogging of the capillary or malfunctioning of the pressure system.

Test 2: Recording a breakthrough curve with the injection pressure

This measurement is also carried out under the above conditions. The time for the appearance of the signal is substantially longer in most CE instruments because the injection pressure is lower than the pressure used to flush the capillary. For troubleshooting, the time required for the breakthrough is most important.

In a third test, parameters must be established that permit checking the potential source. For this, the detection of positively charged solutes in neutral to alkaline buffers is suitable. The analysis times and the selectivities depend strongly on the pH and the buffer concentration. Therefore, the time at which a peak appears is less important than the fact that the positively charged substance reaches the detector at all.

Test 3: Checking the power supply

The capillary described in the first two tests is flushed with a borate buffer (381 mg of disodium tetraborate decahydrate in 100 mL water, pH *ca.* 9.2) and p-aminopyridine (1 mmol L^{-1}) is pressure-injected (3 s). The separation is performed with normal polarity of the power supply of 20 kV; the current should not exceed 30 μA. A large peak appears after less than 10 min.

The characteristic data (two breakthrough times at different pressures, height of the step signal, the current flow, and the appearance of peaks of p-aminopyridine in the borate buffer) should be kept with the instrument documents. A simple comparison of the characteristic data measured before and after the problem saves much time in most cases and compensates for the experience that can only be acquired through long familiarity with CE. It would be desirable to store the test capillary as such and not use it until the "emergency" arises.

11.2 Scenarios of Problems: "What to do if..."

a) Drifts in the migration times

In serial analyses the peak times increase or decrease steadily. This frequently observed phenomenon can have various causes.

(a1) Sample matrix components
Positively charged sample components (analyte or matrix) can block the silanol groups that are responsible for the EOF. In many cases the capillary surface can be regenerated again by flushing with 0.1 M NaOH. The separation method should allow for flushing with sodium hydroxide between analyses. A flush time of 2 to 5 min is sufficient in most cases. It is important that the NaOH not be flushed into the outlet buffer to keep its pH unchanged. A pH gradient would be generated in the capillary for each analysis, which would certainly reduce the reproducibility. This method cannot be used for coated capillaries on account of their low stability towards alkaline solutions.

(a2) The capillary cross-section becomes smaller at one place.
Gradual clogging of a capillary can occur through precipitation of the sample components, through particulates from the sample, or the buffer crystallizing out at the ends. Here, too, flushing with NaOH is effective. NaOH absorbs at 200 nm and can be detected as a step signal. For capillary dimensions similar to those of the test capillary, the breakthrough time should be comparable to that of Test 1. A partially clogged capillary can usually be regenerated by flushing.

(a3) The buffer changes significantly during a series of analyses
At low buffer capacities the pH of the separation buffer changes after only a few analyses. This affects the EOF and the mobility. The sample components that are protonated or deprotonated in the corresponding pH range are especially affected. In critical cases, a pH fluctuation of a few tenths of a unit can decidedly alter the selectivity and, hence, the resolution.

In cases where the buffer capacity is low, changes in pH caused by electrolysis in the buffer vials becomes evident after only 10 to 30 minutes. Fig. 11-1 shows the pH changes of phosphate buffers of different pH values after 5 hours of use. It can be seen clearly that phosphate buffers are suitable for a reproducible series of analyses only at pH 3 and 7, where the capacity is highest. At all of the other pH values significant alterations in pH, and hence EOF instability, can be anticipated. For reproducible work at other pH values, different buffers with high capacity are required, or the buffer should be replenished in *both* vials after each analysis.

The usual analysis times and resolution should be restored with new buffer solutions, and for serial analyses the buffers must be changed correspondingly more often.

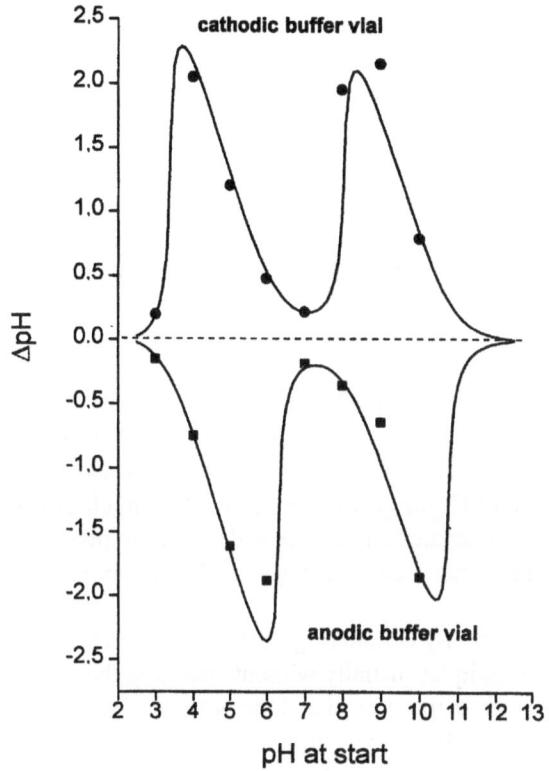

Fig. 11-1
pH changes occurring in the
buffer vials due to electrolysis

Conditions: 10 mM phosphate
buffers of different pH values.
Volume 5 mL, current 30 µA.
The pH was measured after 5
hours of electrophoresis

(a4) Degraded coating
Capillary wall coatings have limited stability (*cf.* chapter on Protein Analysis); their
hydrolysis liberates the active silanol groups that raise the EOF and shorten analysis
times (outlet grounded and EOF toward the cathode). Frequent capillary changes are
very expensive, and analysis conditions should be selected that extend the capillary
lifetime. Usually the degradation of the coating leads to changes in the migration
times and considerable distortion of the peak shape.

(a5) The liquid levels in the inlet and outlet containers differ increasingly
Per analysis, an appreciable amount of buffer is pumped by the EOF from one electrode container to the other. Hydrodynamic pressure acts increasingly against the
EOF, resulting in a drift in the migration times. This problem can be solved by
leveling the liquid heights.

b) No peaks are visible

(b1) Is the sample still okay?
The capillary is filled with buffer and subsequently flushed with the sample solution. This must give rise to a step signal. If a signal is obtained, the detector, the pressure system, and the sample itself can be eliminated as sources of the problem. The separation parameters should then be checked (*cf.* a3). Frequent causes are the polarity of the potential source, the separation potential itself, the position and contents of the buffer, flush, and sample vials. The cause can be ascertained more precisely with the test separation (Test 3).

(b2) Is the capillary clogged? (cf. a2)
If no signal is observed, the capillary can be shortened by about 2 mm at both the inlet and outlet. If still no signal is observed on flushing with NaOH, the capillary must be removed. Capillaries can be flushed with a higher pressure by means of a syringe and an adapter or a connection to an HPLC pump. In most cases this much effort would not be justified by the low cost of uncoated capillaries and it is more practical to install a new one. The absence of the signal may be due to a plugged capillary or inadequate flushing pressure.

The latter can be ascertained definitively by conducting Test 1 with the capillary which is not otherwise used further. It is simpler (usually without changing the capillary) to place the inlet of the capillary in an empty vessel and the outlet in a container with buffer or water. If a capillary is unblocked, air is driven into the outlet by the flushing pressure and bubbles can be seen. Moreover, a step signal can be observed in the detector as the refractive index in the capillary is completely changed as soon as air replaces the buffer in the bore. If a step signal is observed and the air bubbles are visible as well, the search should be continued via the separation parameters. Should only the air bubbles be observed without a signal, the detection unit must be checked further.

(b3) Is the detector still functioning properly?
The verification is performed as described in Test 1, always with respect to the test capillary. This excludes the possibility of other causes (sample or capillary) invalidating the test results. In the absence of a signal, it is practical to start the search by focusing on the capillary itself. For example, the cause may be an unintentional displacement of the detection window so that the polyimide coating hinders detection. Also the splintering off of quartz parts at the detection window (especially if the capillary is held in place with metal parts) can be readily recognized under a magnifying glass or microscope. Some instrument manufacturers have provided checks for wavelength accuracy or lamp energy, but this rarely helps in the problem search. Usually particles or solution residues are in the light path (e.g., after the capillary had been broken at the detection window or the cooling liquid escaped from the cartridge and

moistened parts involved in the detection). A replacement of damaged parts (filter, etc.) or cleaning usually solves the problem.

(b4) Checking the separation parameters

If the capillary is not blocked and the detector functions properly as per Test 1, the source may lie in the separation parameters. Typical examples, as already mentioned, are incorrect polarity of the power supply, wrong detection wavelength, improper positioning of vials by the autosampler control, and incorrect contents in the vials. The power supply can, at least, be checked with Test 3, but further troubleshooting is too specific to detail here. The autosampler control is checked most simply by following the vials (or position in the autosampler) during an analysis to see which are at the inlet and outlet of the capillary at the appropriate times.

(b5) Is the sample injected at all?

Usually, this can be readily answered because two independent injection systems are available in many CE systems. If, for example, pressure injection fails, the sample can still be injected electrokinetically.

Whether electrokinetic injection is proceeding "normally" or is problem-ridden is easily gauged by observing the potential and current courses during the injection. To check the hydrodynamic injection one must be certain that the capillary is not clogged. Test 2 provides information on the proper functioning of the pressure reducer to generate the (usually small) injection pressure.

c) Problems with the baseline

(c1) Drifts

Baseline drift can have several causes and is therefore difficult to correct. For one, sample components may migrate slowly past the detector; for another, electric fields may cause such interference.

A cause that is readily remedied is impurities in the buffer container. Such baseline drift can be eliminated with a new buffer and flushing the capillary with NaOH. Depending on the buffer, the equilibration time (constant analysis time and straight baseline) may range from a few minutes to several hours. Based on our experience, no stable baselines can be obtained with some buffer systems, especially those with stabilization components.

(c2) Spikes

Spikes are extremely short-term fluctuations in the baseline, and their "peak widths" amount to only a few data points (less than a second). They are generally caused by particles in the buffer or sample. These particles are frequently formed as a result of microbial attacks on the buffers since many of these are good growing media for various bacteria. Particularly susceptible are citrate, CAPS, and phosphate buffers of acid and neutral pH. Buffers that have been attacked by microorganisms should not be re-used owing to poor reproducibility, even if the microorganisms can be readily removed by filtration (0.4 μm filter). Filtration of a freshly prepared buffer reduces the bacteria count from the beginning and extends its lifetime. In contrast to the buffers mentioned above, borate buffers, for example, can be stored and used for a very long time.

Besides microorganisms, insoluble buffer components may also give rise to spikes. Large amounts of SDS, for example, dissolve in the common buffers at room temperature, but precipitate at lower temperatures (e.g., in a thermostated auto-sampler) due to the rapid decrease in its solubility.

(c3) Is the noise unusually large?
The reason for a high noise level is usually too little light reaching the photosensitive layer. If UV absorbing buffer systems are used, they should not be too concentrated to avoid high baseline noise. This may also be caused by too low a light intensity from the lamp. The lamp energy should be checked regularly if the instrument offers this option, so that it can be replaced before significant losses in sensitivity appear.

(c4) Baseline shifts
Displacements in the baseline are called shifts. In some CE instruments shifts are observed primarily within the first minute of a separation. They can be caused by electric fields that perturb the electronics of the detector. The shifts usually appear very reproducibly and can be suppressed by autozeroing. If they appear later, discontinuities in the buffer system (or for large injection volumes, in the sample itself) may be responsible. Renewal of the buffer in both containers is usually beneficial and should be performed at regular intervals when handling larger series of samples.

12 LITERATURE INDEX

12.1 Literature Cited

[1]   S. Hjerten, Chromatogr. Rev., 9 (1967), 122
[2]   R. Virtanen, Acta Polytech. Scand., 123 (1974), 1.
[3]   F.E.P. Mikkers, F.M. Everaerts, T.P.E.M. Verheggen, J. Chromatogr. 169 (1979), 11
[4]   J.W. Jorgenson, K.D. Lukacs, J. Chromatogr. 218 (1981), 209
[5]   J.W. Jorgenson, K.D. Lukacs, Anal. Chem. 53 (1981), 1298.
[6]   H. Engelhardt, W. Beck, J. Kohr, T. Schmitt, Angew. Chem. 105 (1993), 659
[7]   J. Pospíchal, P. Gebauer, P. Bocek, Chem. Rev. 89 (1989), 419
[8]   A.W. Adamson, Physical Chemistry of Surfaces (4th Ed.), Wiley & Son, New York (1982)
[9]   J. Kohr, H. Engelhardt, J. Microcol. Sep. 3 (1991), 491
[10]   B.B. VanOrman, G.G. Liversidge, G.L. McIntire, T.M. Olefirowicz, A.G. Ewing, J. Microcol. Sep. 2 (1990), 176
[11]   I. Halasz, Z. Anal. Chem. 277 (1975), 257
[12]   J.C. Giddings, Unified Separation Science, Wiley and sons, New York (1991)
[13]   J.H. Knox, Chromatographia 26 (1988), 329
[14]   D. McManigall, S.A. Swedberg, Tech. in Protein Chem., T. Hugli ed., Academic Press: San Diego (1989), 468
[15]   W. Beck, Nachr. Chem. Techn., 41 (1993) March
[16]   J. Kohr, H. Engelhardt, J. Chromatogr. A, 652 (1993), 309
[17]   X. Huang, T. Pang, M. Gordon, R. Zare, Anal. Chem. 59 (1987), 2747
[18]   X. Huang, M.J. Gordon, R.N. Zare, Anal. Chem. 60 (1988), 375
[19]   M. Deml, F. Foret, P. Bocek, J. Chromatogr. 59 (1985), 320
[20]   R.L. Chen, D.S. Burgi, J. Chromatogr. 559 (1991), 141
[21]   T.J. Thompson, F. Foret, P. Vourous, B.L. Karger, Anal. Chem. 65 (1993), 900
[22]   G. Bruin, G. Stegeman, A. Van Asten, X. Xu, J. Kraak, H. Poppe, J. Chromatogr. 559 (1991), 163
[23]   M. Fuchs, P. Timmoney, M. Merion, Poster No. PM-24, HPCE 1991, San Diego, CA
[24]   J.P. Chervet, United States Patent 5,057,216, October 15 (1991)
[24a]   R. C. Williams, J. F. Edwards, C. R. Ainsworth, Chromatographia 38 (1994), 441
[25]   G.B. Gordon, United States Patent 5,061,361, October 29 (1991)
[25a]   C. Chiesa, C. Horváth, J. Chromatogr. 645, (1993), 337
[26]   C. Haber, I. Silvestri, S. Röösli, W. Simon, CHIMIA,1991, 45, 117
[27]   R. A. Wallingford, A. G. Ewing, Anal. Chem. 59 (1987), 678
[28]   W.M.A. Niessen, J. van der Greef, Liquid Chromatography-Mass Spectrometry (Chromatographic Science Series, Vol. 58) Marcel Dekker (1992), R. 273ff
[29]   J. Liu, Y.-Z. Hsieh, D. Wiesler, M. Novotny, Anal. Chem. 63 (1991), 408
[29a]   N. J. Reinhoud, U. R. Tjaden, J. van der Greef, J. Chromatogr. A 673, (1994), 255
[30]   M. Albin, R. Weinberger, W. Sapp, S. Moring, Anal. Chem. 63 (1991), 417
[31]   D.H. Patterson, B.J. Harmon, F.E. Regnier, J. Chromatogr. A 662 (1994), 389
[32]   W. Beck, H. Engelhardt, Chromatographia, 33 (1992), 313

[33] K.D. Altria, R.C. Harden, M. Hart, J. Hevizi, P.A. Hailey, J.V. Makwana, M.J. Portsmouth J. Chromatogr. 641 (1993), 147
[34] J.C. Giddings, Unified Separation Science, John Wiley & Sons, New York (1991)
[35] W. Beck, R. van Hoek, H. Engelhardt, Electrophoresis, 14 (1993), 540
[36] J. Romano, P. Jandik, W.R. Jones, P. E. Jackson, J. Chromatogr., 546 (1991), 411
[37] H. Small, T. Miller, Anal. Chem., 54 (1982), 463
[38] F. Foret, S. Fanali, A. Nardi, P. Bocek, Electrophoresis, 11 (1990), 780
[39] W. Beck, H. Engelhardt, Fresenius J. Anal. Chem. 346 (1993), 618
[40] W. Beck, H. Engelhardt, Fortschrittsberichte 1992 (Würzburger Kolloqium) J.H. Schneider, Bertsch Verlag, p. 27
[41] F.-T.A. Chen, J. Chromatogr. 559 (1991), 445
[42] H.H. Lauer, D. McManigill, Anal. Chem. 58 (1986), 166
[43] J.S. Green, J.W. Jorgenson, J. Chromatogr. 478 (1989), 63
[44] M. Bushey, J. Jorgenson, J. Chromatogr. 480 (1989), 301
[45] J.A. Bullock, L.C. Yuan, J. Microcol. Sep. 3 (1991), 241
[46] Waters Application Note (1991).
[47] J.K. Towns, F.E. Regnier, Anal. Chem. 63 (1991), 1126
[48] A. Emmer, M. Jansson, J. Roerade, J. Chromatogr. 547 (1991), 544
[49] E. Kenndler, K. Schmidt-Beiwl, J. Chromatogr. 545 (1991), 397
[50] J. Jorgenson, K. Lukacs, Science 222 (1983), 266
[51] A. Balchumas, M. Sepaniak, Anal. Chem. 59 (1987), 1466
[52] A. Dougherty, C. Woolley, D. Williams, D. Swaile, R. Cole, M. Sepaniak, J. Liq. Chromatogr. 14 (1991), 907
[53] M. Sepaniak, D. Swaile, A. Powell, R. Cole, HRC & CC 13 (1990), 679
[54] G.Bruin, J. Chang, R. Kuhlman, K. Zegers, J. Kraak, H. Poppe,J. Chromatogr. 471 (1989), 429
[55] G. Bruin, R. Huisden, J. Kraak, H. Poppe , J. Chromatogr. 480 (1989), 339
[56] R. McCormick, Anal. Chem. 60 (1988), 2322
[57] S. Swedberg, Anal. Biochem. 185 (1990), 51
[58] Y.-F. Maa, K. Hyver, S. Swedberg, HRC & CC 14 (1991), 65
[59] J. Kohr, H. Engelhardt, J. Microcol. Sep. 3 (1991), 491
[60] S. Hjerten; J. Chromatogr. 347 (1985), 191
[61] S. Hjerten, K. Ellenbring, F. Kilar und J. Liao; J. Chromatogr. 403 (1987), 47
[62] K.Cobb, V. Dolnik, M. Novotny, Anal. Chem. 62 (1990), 2478
[63] M. Strege, A. Lagu, Anal. Chem. 63 (1991), 1233
[64] C. Bolger, M. Zhu, R. Rodriguez, T. Wehr, J. Liq. Chromatogr. 14 (1991), 895
[65] F. Regnier, J. Towns, J. Chromatogr. 516 (1990), 69
[66] D. Bentrop, J. Kohr. H. Engelhardt, Chromatographia 32 (1991), 171
[67] S.Terabe, H. Utsumi, K. Otsuka, T. Ando, T. Inonata, S. Kuze, Y. Hanaoka, HRC & CC 9 (1986), 666
[68] J. Lux, H. Yin, G. Schomburg, HRC & CC 13 (1990), 145
[69] S. Mayer, V. Schurig, HRC & CC 15 (1992), 129
[70] D.W. Armstrong, Y. Tang, T. Ward, M. Nichols, Anal. Chem. 65 (1993), 1114
[71] G. Guiochon, C.L. Guillemin, J. Chromatogr. Library 4, Quantitative Gas Chromatography, Elsevier (1988)
[72] M. Gilges, M. H. Kleemiß, G. Schomburg, Anal. Chem. 66 (1994), 20
[73] S. Terabe, K. Otsuka, K. Ichikawa, A. Tsuchiya, T. Ando, Anal. Chem. 56 (1984), 111
[74] J. Vindevogel, P. Sandra in: Chromatographic Methods, Introduction to Micellar Electrokinetic Chromatography, Hüthig Verlag, Heidelberg (1992)

[75]   S. Terabe, Trends Anal. Chem. 8 (1989), 129
[76]   S. Terabe, K. Otsuka, T. Ando, Anal. Chem. 57 (1985), 834
[77]   S. Terabe in: Norberto A. Guzman, Capillary Electrophoresis Technology, Chromatographic Science Series, 64, Marcel Dekker, New York (1993), p. 65
[78]   H.T. Rassmussen, H.M. McNair,. HRC&CC. 12 (1989), 635
[79]   S. Terabe, Oral Presentation, 6 th International Symposium on HPCE, San Diego (1994)
[80]   S. Terabe, J. Chromatogr. 545 (1991), 359
[80a] Beckman Chromatogram, Aug. 1990
[81]   S. Fanali, J. Chromatogr. 474 (1989), 441
[82]   A. Guttman, A. Paulus, A.S. Cohen, N. Grinberg, B.L. Karger, J. Chromatogr. 448 (1988), 41
[83]   S. Fanali, P. Bocek, Electrophoresis 11 (1990), 757
[84]   J. Snopek, H. Soini, M. Novotony, E. Smolkova-Keulemansova, I. Jelinek, J. Chromatogr. 559 (1991), 215
[85]   T. Schmitt, H. Engelhardt, HRC & CC 16 (1993), 35
[85a] T. Schmitt, H. Engelhardt, Chromatographia 37 (1993), 475
[86]   M. Heumann, G. Blaschke, J. Chromatogr. 648, (1993), 267
[87]   S. Terabe, H. Ozaki, K. Otsuka, T. Ando, J. Chromatogr. 332 (1985), 211
[88]   N.W. Smith, J. Chromatogr. A 652 (1993), 259
[89]   R.R. Rajewski, V.J. Stella, US Patent (pending) 07/469.087 (1992]
[90]   S. Terabe, Trends Anal. Chem. 8 (1989), 129
[91]   A. Nardi, A. Eliseev, P. Bocek, S. Fanali, J. Chromatogr. 638 (1993), 247
[92]   K. Otsuka, S. Terabe, Chromatogr. 515 (1990), 221
[93]   K. Otsuka, S. Terabe, Electrophoresis 11 (1990), 982
[94]   J. Mazzeo, E. Grover, M. Swartz, J. Peterson, Waters Chromatography Division of Millipore (unpublished results)
[95]   K. Otsuka, J. Kawahara, K. Takekawa, S. Terabe, J. Chromatogr. 559 (1991), 209
[96]   A. Dobashi, T. Ono, S. Hara, J. Yamaguchi, Anal. Chem., 61 (1989), 1984
[97]   A. Dobashi, T. Ono, S. Hara, J. Yamagushi, J. Chromatogr. 480 (1989), 413
[98]   S. Terabe, M. Shibata, Y. Miyashita, J. Chromatogr. 480 (1989), 403
[99]   H. Nishi, T. Jukuyama, M. Matsuo, S. Terabe, J. Microcolumn Separation 1 (1989), 234
[100] H. Nishi, T. Fukuyama, M. Matsuo, S. Terabe. J. Chromatogr. 515 (1990), 233
[101] E. Gassmann, J.E. Kuo, R.N. Zare, Science, 230 (1985), 813
[102] P. Gozel, E. Gassman, H. Michelsen, R.N. Zare, Anal. Chem. 59 (1987), 44
[103] R. Kuhn, F. Stoecklin, F. Erni, Chromatographia 33, (1992), 32-36.
[104] R. Kuhn, F. Erni, T. Bereuter, J. Hauser, Anal. Chem. 64 (1992), 2815
[105] S. Fanali, L. Ossicini, F. Foret, P. Bocek, J. Microcolumn Separation 1 (1989), 190
[106] S. Birnbaum, S. Nillson, Anal. Chem. 64 (1992), 2872
[107] P. Sun, G.E. Barker, R.A. Hartwick, N. Grinberg, R. Kaliszan, J. Chromatogr. A 652 (1993), 247
[108] L. Valtcheva, J. Mohammad, G. Petterson, S. Hjerten, J. Chromatogr. 638 (1993), 263
[109] D. Belder, G. Schomburg, HRC&CC 15 (1992), 686
[110] S.A.C. Wren, J. Chromatography, J. Chromatogr. 636 (1993), 57
[111] Y.Y. Rawjee, D.U. Staerk, G. Vigh, J. Chromatogr. 635 (1993), 291
[112] A.S. Cohen, D.R. Najarian, B.L. Karger, J. Chromatogr. 516 (1990), 49
[113] M. Strege, A. Lagu, Anal. Chem. 63 (1991), 1233
[114] D.N. Heiger, A.S Cohen, B.L Karger, J. Chromatogr. 516 (1990), 33
[115] Y.F. Pariat, J. Berka, D.N. Heiger, T. Schmitt, M. Vilenchik, A.S. Cohen, F. Foret, B.L. Karger, J. Chromatogr. A 652 (1993), 57
[116] M.C. Ruiz-Martinez, I. Smirnov, J. Berka, F. Foret, B.L. Karger, (private communication)

[117] K. Ganzler, K.S. Greve, A.S. Cohen, B.L. Karger, A. Guttman, N.C. Cooke, Anal. Chem 64 (1992), 2665

[118] P. Serwer, Electrophoresis 4 (1983), 375

[119] G.W. Slater, J. Noolandi, in L. Lee (Ed.), New Trends in Physics and Physical Chemistry of Polymers, Plenum, New York (1989), 547

[120] D.L. Smisek, D.A. Hoagland, Macromolecules 22 (1989), 2270

[121] D;. Schwartz, M. Koval, Nature 338 (1989), 520

[122] P.G. Righetti, Separation and Purification Methods (1975), 23.

[123] P.C. Righetti, Isoelectric Focusing: Theory, Methodology and Applications, Elsevier, Amster dam (1983)

[124] B.J. Radola in Isoelectric Focusing: (Hrsg.: N. Catsimpoolas), Academic Press, New York (1976), S. 119 f

[125] P.G. Righetti, Immobilized pH Gradients: Theory and Methodology, Elsevier, Amsterdam (1990)

[126] H. Swensson, Prot. Biol. Fluids, 15 (1967), 515

[127] J.R. Mazzeo, I.S. Krull, Anal. Chem. 63 (1992), 2852

[128] P.G. Righetti, G. Tudor, K. Ek, J. Chromatogr. 220 (1981), 115

[129] P. Bocek, M Deml, P. Gebauer, V. Dolnik; Analytical Isotachophoresis, B.J. Radola, Ed., VCH Publishers, New York (1988)

[130] 7 th Int. Symp. on CE and ITP, J. Chromatogr. 545 (1991), 2 ff.

[131] J. Caslavska, S. Lienhard, W. Thormann, J. Chromatogr. 638 (1993), 335

[132] J.H. Knox, I.H. Grant, Chromatographia 32 (1991), 317

[133] H. Yamamoto, J. Bauman, F. Erni, J. Chromatogr. 593 (1992), 313

[134] H. Engelhardt, High Performance Liquid Chromatography (int. ed. in engl.) Springer Verlag (1979)

12.2 Additional Literature Sources

Reviews on the subject of "capillary electrophoresis"

J.W. Jorgenson, K.D. Lukacs; Science 222 (1983), 266

J. W. Jorgonson, Anal. Chem. 556 (1986), 743A

M.J. Gordon, X. Huang, S.L. Pentoney, R.N. Zare; Science 242 (1988), 224

A. Ewing, R. Wallingford, T. Olefirowicz, Anal. Chem. 61 (1989), 292A

B.L. Karger, A.S. Cohen, A. Guttman; J. Chromatogr. (Biomed. App.) 492 (1989), 585

A.G. Ewing, R.A. Wallingford, T.M. Olefirowicz; Anal. Chem. 61 (1989), 292A

N.A. Guzman, L. Hernandez, B.G. Hoebel, BioPharm, 1989, Jan.

D.M. Goodall, D.K. Lloyd, S.J. Williams, LC/GX Int. 3 (1990), 28

W.G. Kuhr, Anal. Chem. 62 (1990), 403R

P.D. Grossman, H.H. Lauer, S.E. Moring, D.E. Mead, M.F. Oldham, J.H. Nickel, - J.R.P. Goudberg, A. Krever, D.H. Ransom, J.C. Colburn, Amer. Bio. Lab. 1990, Feb., 35

E.S. Yeung, W.G. Kuhr; Anal. Chem., 63 (1991), 275

H.H. Lauer, J.B. Ooms, Analytica Chimica Acta 250 (1991), 45

G. Schomburg, Trends Anal. Chem, 10 (1991), 163

R. Mazzeo, I.S. Krull, BioTechniques 10(5) (1991), 638

D. M. Goodall, S. J. Williams, D. K. Loyd, Trends Anal. Chem. 10 (1991), 272

W.G. Kuhr, C.A. Monnig, Anal. Chem. 642 (1992), 389R

R. Kuhn, S. Hoffstetter-Kuhn, Chromatographia 34, (1992), 505

H. Engelhardt, J. Kohr, W. Beck, T. Schmitt, Angew. Chem. Int. Ed. Engl. 32 (1993), 629

Books on the subject of "capillary electrophoresis"

S.F.Y. Li: „Capillary Electrophoresis", J. Chromatogr. Libr. 52, Elsevier Amsterdam (1992)

J. Vindevogel, P. Sandra in: Chromatographic Methods, „Introduction to Micellar Electrokinetic Chromatography", Hüthig Verlag, Heidelberg (1992)

P. Jandik, G.K. Bonn: Capillary Electrophoresis of small molecules and ions, VCH Publishers, Weinheim/New York (1993)

R. Kuhn, S. Hoffstetter-Kuhn, „Capillary Electrophoresis, Principle and Practice", Springer-Verlag Berlin Heidelberg (1993)

R. Weinberger, „Practical Capillary Electrophoresis", Academic Press (1993)

N. A. Guzman, „Capillary Electrophoresis Technology", Chromatographic Science Series, 64, Marcel Dekker, INC. (1993)

F. Foret, L. Krivankova, P. Bocek, „Capillary Zone Electrophoresis", B.J. Radola (Ed.), VCH (1993)

R.A. Wallingford, A.G. Ewing, „Capillary Electrophoresis"; Advances in Chromatography, Vol. 29 (Ed. J.C. Giddings, E. Gruschka, P.R. Brown), M. Dekker, Inc. (1989)

D.N. Heiger, „High Performance Capillary Electrophoresis", An Introduction, Hewlett Packard (Waldbronn, Germany 1993)

P.D. Grossman, J.C. Colburn eds., „Capillary Electrophoresis - Theory and Practice", Academic Press Inc.: San Diego (1992)

Symposium volumes

1st International Symposium on High Performance Capillary Electrophoresis (Boston), J. Chromatogr. 480 (1989)

2nd International Symposium on High Performance Capillary Electrophoresis (San Francisco), J. Chromatogr. 516 (1990)

3rd International Symposium on High Performance Capillary Electrophoresis (San Diego), J. Chromatogr. 559 (1991)

4th International Symposium on High Performance Capillary Electrophoresis (Amsterdam), J. Chromatogr. 608 (1992)

5th International Symposium on High Performance Capillary Electrophoresis (Orlando), J. Chromatogr. 652 (1993)

6th International Symposium on High Performance Capillary Electrophoresis (San Diego), J. Chromatogr. 680 (1994)

7th International Symposium on High Performance Capillary Electrophoresis (Würzburg), J. Chromatogr. 716 and 717 (1995)

13 ACKNOWLEDGEMENT

We express our sincere appreciation to the following for permission to reproduce the figures used in this book:

Friedrich Vieweg & Sohn Verlagsgesellschaft mbH, Wiesbaden
Elsevier Science Publishers B.V., Amsterdam
American Chemical Society, Washington
Marcel Dekker, INC., New York
VCH Verlagsgesellschaft mbH, Weinheim
Hüthig Buch Verlag, Heidelberg
Dr. Milton L. Lee, Provo, Utah, USA

The following figures were redroduced with permission:

Figure: 2-1, 2-2, 3-8, 3-9, 4-14, 5-2, 5-20, 6-11, 6-12, 6-13 from [6]
Figure: 3-4 und 5-35 from [9]
Figure 3-5, 3-6 from [10]
Figure: 4-4 from [18]
Table 4-5 from [18]
Figure: 4-6 from [20]
Figure: 4-9 from [22]
Table: 4-10 from [33]
Figure: 4-12 from [24a]
Figure: 4-16 from [25a]
Figure: 4-19, 4-20 from [29a]
Figure: 4-21, 5-11, 5-12, 5-15 from [32]
Figure: 5-7 from [35]
Figure: 5-13 from [40]
Figure: 5-28 from [41]
Figure: 5-31, 5-32 from [52]
Figure 5-33 from [57]
Figure: 5-34 from [54]

Figure: 5-36 from [59]
Figure 5-38 from [65]
Figure 5-39 von [72]
Figure 5-40 from [66]
Figure 6-2 from [75]
Figure: 6-6 from [73]
Figure: 6-1, 6-5 from [74]
Figure: 6-7 from [78]
Figure: 6-8 from [80]
Figure 6-9 from [80a]
Figure: 7-16 from [85]
Figure 7-25 von [94]
Figure: 7-26 from [100]
Figure 8-8 von [116]
Figure 8-9 from [117]
Figure 8-11 from [118]
Figure 8-12 from [119]
Figure 10-2 from [131]
Figure 10-3 from [133]

SUBJECT INDEX